T R A N S M O N T

MW00737602

Stranger Wycott's Place

Published by New Star Books

Series Editor: Terry Glavin

Other books in the Transmontanus series

Stranger Wycott's Place

STORIES FROM THE CARIBOO–CHILCOTIN

John Schreiber

TRANSMONTANUS ׀ NEW STAR BOOKS VANCOUVER

Stranger Wycott's place up Churn Creek above Wycott Flats.

CONTENTS

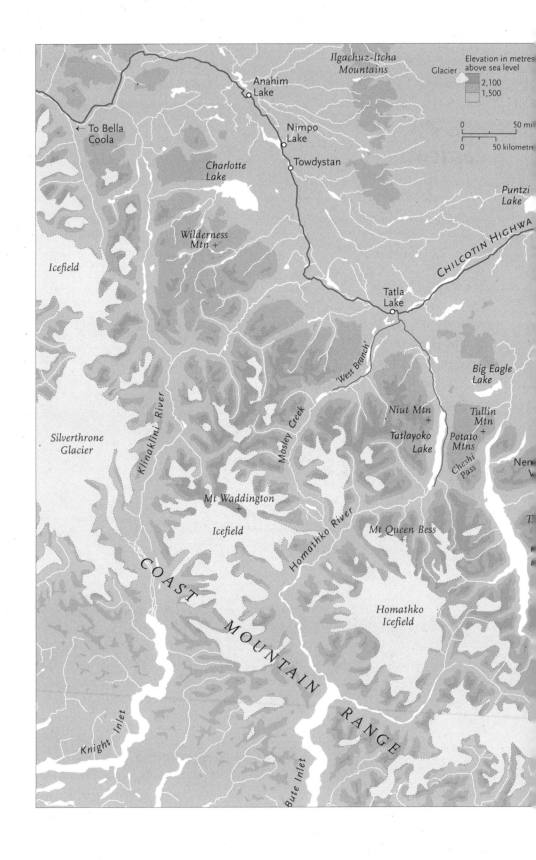

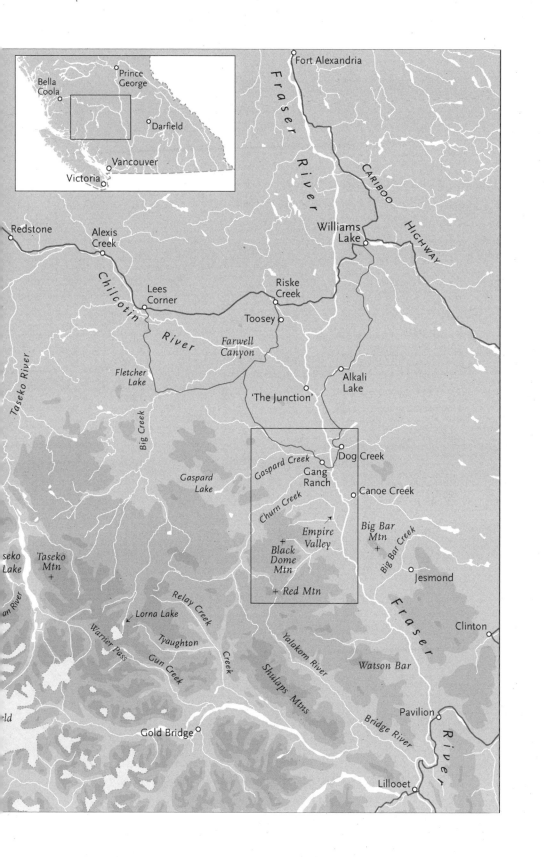

Map on previous spread shows the Chilcotin plateau and the west Cariboo. See p. 82 for a detailed map of the area around Gang Ranch (inset map, lower right).

For my parents, Alizon and Pat Schreiber,
who taught us to love this land we are a part of;
for Marne who has shown me that love is
unconditional and without limits.

Established in Yoga, O winner of wealth,
Perform actions ...
    Bhagavad Gita
    Chapter Two, verse 48

ONE

Beginnings: Dry Air and Trembling Poplar Leaves

There is a dream dreaming us.

— KALAHARI BUSHMAN HUNTER
[TO LAURENS VAN DER POST]

Don Brooks and I were sitting on his deck in the late afternoon sunlight and drinking something red and excellent from Blue Mountain Vineyard. We had just returned from an extended walking trip in the South Chilcotin Mountains and were basking in its afterglow, listening to the long, slow, excruciating, descending chords and measured bell beats of Arvo Pärt's *Cantus in memoriam Benjamin Britten*. Don said, "You should write stories like that."

"That's what I'm trying to do," I replied, and to myself, silently, fervently, "help me make them real."

These stories, and many more, written or waiting to be written, are most obviously about the Cariboo–Chilcotin region of south-central British Columbia and the great Fraser River that bisects this vast land, and about wildness and walking, mountains and old trails, coyotes and bighorn sheep, aboriginal folks, homesteaders, ranchers and history. Living these stories, and thinking and writing about them, has sharpened my focus, most thoroughly, on the fundamental fact of this place where we are, this homeland we live in. In so doing, I experience a growing sense that the living world is even more lively than I had thought possible. Through learning where we are, we may learn who we are.

As a young boy after the Second World War, I lived with my

family up the North Thompson River north of Kamloops near
a post office stop on the highway called Darfield. Our partially
built log home was up a long hillside road above the valley, alone
but for the Schilling family a quarter mile past us. There, for several years, my father attempted to make a living for us as a guide-
outfitter to would-be trout fishermen and hunters of moose and
mule deer. We subsisted well enough on deer and moose meat,
macaroni, porridge, home-cured bacon, and vegetables in season
or out of cans and Mason jars. But the marginal income drove
us out four years later. We moved to Port McNeill, a large, family logging camp on the north end of Vancouver Island, where
Dad could resume his pre-war occupation and we could make a
decent livelihood.

I spent much of my time up there at Darfield on my own, playing
around the lodge or walking down and up the hillside to school in
the valley, a two- to three-mile jaunt each way, and over to the
Schillings' farm for milk, often in the semi-dark of early evening.
Sometimes I would go over there to play with the younger Schilling kids, all older than me, who, as I look back, must have been
kind and tolerant young people to have put up so regularly with
this small ragamuffin out of the bush, freshly arrived at their door-
step. I keep up with these folks to this day.

So I imprinted to my own walking, to wildness, and most especially to those classic markers of the southern BC Interior: dry air,
redolent "bee-loud" wild-rose hillsides, open woods, dry-belt fir
trees, cool nights and clear night skies, golden autumns, and the
heart-shaped leaves of quaking aspens fluttering in the slightest
breezes. I learned in my early growing up to be quite at ease with
my alone self, with the untamed shadowy woods, the dark, and
the usually sudden sightings or sounds of creatures, mostly wild:
coyotes, owls, deer, porcupines, quick weasels in the woodpile; a
snake with a toad feet-first down its distended throat; a tiny, pug-
nacious, sharp-toothed shrew crossing the road; a startled hawk
on a branch over the outhouse in the dim light of evening; Mr.
Jannings' cattle as I walked through them on the way to school;
a wounded black bear up the trail to Frog Lake; our turned-out
horses moving around in the night; our gone-wild cats returning
to be fed in winter; a moose under my window; wolves howling at
twilight in the bright snowy hills above the lodge.

All this left a mark on me. Those knowledgeable about child development tell us we attach to where we live and play when we are seven and eight years of age. In my adult life I have felt an underlying need to try to define and express that old mysterious sense from my child years: my memories of walking through landscape alone; my familiarity with this home ground of ours; these diverse BC places, people, animals; and, always, those characteristic aspects of the southern Interior parkland country. These memories are at times distinct, at times dim, but they are, at all times, deep.

My growing inclination is to pay attention to these memories of place and to the places where I am now as closely and as often as I can. Such attention leads me to experience and acknowledge more thoroughly the richness and liveliness of detail, the power of the intrinsic truth of things and the subtlety of existence. I sense that, at root, meaning begins with the ground we stand and walk on and are so absolutely dependent upon. It begins with the connections and relationships that ensue with and from those countless, various places.

As I have driven and walked through some portions of the British Columbia backcountry over the years, I have come, inevitably, to wonder more deeply about wilderness and wildness. What makes a place wild? Is wilderness only where humans are absent and our effects minimal? Does "wild" have intrinsic meaning in and of itself, or is it merely defined in relative terms by what it is not? Can places where it feels as if the spirit of wildness has retreated or disappeared recover themselves? Can wild return? Perhaps the quality and modesty of our attention to a place can assist in that process.

And how does myth fit here? Since our beginnings, human cultures, without exception, have expressed themselves in terms of myth, but in our technologically driven, postmodern era we seem unaware of what myth meant in older times, what it might mean to us now and how it could express itself in our beliefs and actions. We have come to see the word "myth" as another word for "lie" or "falsehood." In a parallel way, we have denigrated the natural world around us from something animate and spirit-filled to something to be used and thrown away. Only recently have we begun to recognize again the need to see the organic realm of plants and animals as fully alive as ourselves and worthy of our acknowledgement and respectful attention. Perhaps we might begin to see the less-obviously organic world of swamps, rivers and oceans, and even rocks and mountains, in a comparably honourable manner. As recent settlers on this North American, western Canadian land, can we come to see ourselves in terms of myth — the old land-based myths that were here before our arrival and the post-contact myth stories that may be emerging now? What could such myths be? How might we be a part of them? Do we dare ask if myths come to see themselves in us? And what are the connections between myth and wildness? Myth itself may be wild.

Such questioning requires careful, patient attention. We might seek answers as if we were walking on an old trail in the woods past some half-forgotten homestead perhaps, or on a route high up over some exposed ridge. As we walk we are watchful, looking for details and to see what is out of place, looking over our shoulder to note where we have come from, wondering what is around and ahead of us and how we will reach our destination. There will be differences and uncertainties, places we have not seen, side trails we do not know, birdcalls we have not heard before, animal tracks we do not recognize, and changes in the wind, changes in the weather. So we learn to learn slowly, in increments, testing the facts and our deductions, not jumping to fast conclusions and always checking our basic assumptions. Many an exploration has faltered on false assumptions. Sometimes we simply walk with our minds blank, free, our mouths slightly open perhaps, absorbing what there is around and within us. (You might consider reading these stories in a similar manner.) As we look and listen and connect, answers occur, more questions arise and over time we may come to sense more thoroughly that all aspects of land are lively and that stories, even myth stories, can emerge and come alive.

TWO

The River

The river flows on like a breath
in between our life and death;
tell me,   who is the next
to cross   the borderline.

'ACROSS THE BORDERLINE'

RY COODER, JOHN HIATT, DAVE DICKENSON

A trip to the Interior really begins when I cross the Fraser River on the long bridge above the gravel bars, corn fields and cottonwood margins at Agassiz. This time, the corn is tall already, and the river is high and muddy; only the tops of the gravel bars are showing. I take that straight, true, relatively carless highway northeast past Seabird Island, Ruby Creek and Gordon Hanson's archaeological dig at the ancient village site by the foot of the pipeline crossing.

I cannot pass Ruby Creek without thinking of that short letter to the *Vancouver Sun* written some time back, in the late sixties I believe, by a probably elderly woman from that place. There had been a spate of sasquatch sightings, mainly by Indian people up the Fraser Valley, including a number around Ruby Creek, heart of sasquatch land, and, consequently, much discussion with some controversy. This woman responded emphatically. "The white-man doesn't see the sasquatch because he does not have the eyes to see the sasquatch," she wrote.

The trip begins again with the turn of the river and the turn of the highway north, at Hope. I don't usually stop there, but Hope is an interesting place. It has many strong reminders of the old fur trade and goldrush days — such as Christ Church, with its original artifacts and structural details — and of even earlier times, notably the deep kekuli holes, large depressions in the ground where pithouses had been, at Telte Yet Campsite on the river. Especially, there are trails from the old fort site over the Cascade Mountains to the southern Interior: the Hudson's Bay Company Brigade Trail northeast to Tulameen and Fort Kamloops; the Dewdney and Hope Trails to Whipsaw Creek and points east; the Whatcom Trail down the Skagit River and over the border south; and John Fall Allison's route to the upper Similkameen over the pass named after him. I have walked on some of those trails, breathing in the resonances of those who had walked or ridden there before. I have stood on six feet of hard snow at Hope Pass on a sunny, clean spring morning with hiking buddy Trevor Calkins, both of us exhilarated to be there. These routes have been used since the ice age, over 10,000 years ago. Trails are arteries; old, used trails are my curiosity and passion.

Just north of the Hope junction, at Lake of the Woods, formerly Shxwqo:m, a place of power in pre-contact days, there is a small park and parking lot beside the highway, overlooking the lake. At ground level, in the thick bark of one of the Douglas fir trees lining the edge, is a heavy, sap-encrusted, industrial chain, thoroughly embedded, with a broken hook, hanging. On the hook are stamped some worn and mainly indecipherable words, the ones readable being "Toronto" on one side and "Canada" at the end of the other. Was it used as a tie-up for a log boom in the lake when the area was logged a hundred years ago, or is it a remnant from the wagon-hauling days nearly half a century earlier? My guess would be the former.

The trip continues through Yale, the start of the Cariboo Road, with the stone foundations of stores and houses on Front Street, the phantom paddlewheelers lined up along the steep beach below, and the Church of St. John the Divine in view from the highway. As in Hope, and in Spuzzum up the road, the streets of Yale were laid out in grid pattern by the sappers of the Royal Engineers in the latter days of the goldrush. The charming old church, and Christ Church in Hope, both Anglican churches, were built about the same time, 1859 and 1861 respectively, making them among the oldest buildings in British Columbia.

In January 1859, Yale was the setting for much seminally significant history, including the showdown between Judge Matthew Begbie, Colonel Richard Moody and Ned McGowan, the calculating California "judge," gold miner and power seeker. This event became known as "McGowan's War." When McGowan and his mostly American followers got into a squabble with a local magistrate, Governor James Douglas promptly sent Begbie and Moody and a small number of Royal Engineers up the Fraser in the cold and wet of winter to show the British flag and keep the peace. America was in an expansionary mood. Douglas could see McGowan's little insurrection as potential grounds for the extension of U.S. influence north and, stout British subject that he was, acted firmly to stop it. Ned McGowan backed down and left for California soon after. The Yale incident is a much understated turning point in BC and Canadian history. Because of the clear determination and quick actions of Douglas and his officers, our future as Canadians out here on the western edge of the country was preserved, at least to now.

North of Yale the river canyon steepens and deepens, and serious highway and railroad tunnels begin. Traditional trails were precipitous and dangerous, as some of the comments in Simon Fraser's journals attest. He states that the canyon "is so wild that I can-

The site of the paddlewheeler landing below Front Street at Yale, looking upriver.

not find words to describe our situation at times. We have to pass where no human being should venture." The several subsequent whiteman's routes to the goldfields were difficult to construct and maintain. The Engineers' road up the edges of the Fraser gorge, cribbed out into open space in steep places, was something of a miracle. Along this stretch of river runs the boundary between downriver Halkomelem-speaking Sto:lo people and the Interior

The Anglican Church of St. John the Divine at Yale, one of the oldest buildings in British Columbia.

Salish Thompson or Nlaka'pamux, "the people of the canyon," who live along the Fraser past Lytton, and up the Thompson and Nicola Rivers as far as Ashcroft and the Merritt region.

Across the river along here is a patch of wild oaks. With some trees growing on the south side of Sumas Mountain, these are, according to Chess Lyons in *Trees, Shrubs and Flowers to Know in British Columbia*, the only known Garry oaks growing on the Canadian mainland. These are, or were, species on the edge of their ranges, like spotted owls or sage grouse in British Columbia and those seven or eight Oregon ash shrubs in Beacon Hill Park in Victoria, formerly growing wild, I think, but recently and probably heedlessly rooted out. The Okanagan Valley, the next major valley east, is full of such species on the margins, many threatened with extirpation by development and interruption. The lower Fraser Canyon is at the ecological cusp between wet coast and dry interior and has plant and animal varieties common to both regions; such border areas are rich in diversity and are all the more lively for that. Each bend in the river presents subtly varying aspects of the emerging southern BC Interior bio-region.

In the summer of 1946, my family passed along that same old Cariboo Road route in a superannuated flat-deck truck, most of our worldly goods packed behind under a big white tarpaulin. Mother held my recently born brother Chris, a tiny bundle, in her arms; Dad was doing the driving. I was just five and thought that truck with its white "tent" on the back a fabulous thing. We were following the wartime dream of my father, a practised woodsman and hunter, to make a living for his family as a guide-outfitter at Darfield, fifty miles up the North Thompson River from Kamloops. My strongest memories of that first trip to the Interior were frequent glimpses of the Fraser River down in the bottom of its trench, and the narrow, bendy, wood-cribbed, one-lane highway around Jackass Mountain. That name, and an image of a single fir tree at the top of a steep rock bluff, stark against high sky blue, a child's glance into the core of existence, enduring and unchanging, persist in my memory.

For my parents, on the other hand, the trip was probably remembered for concerns about what was to come, and for my repeated

Darfield Lodge, where the Schreiber family lived in the mid to late 1940s. The lodge was built by Pat Schreiber and Kori Norgaard in 1946–47.

question "What lake is that, Daddy?" each time I spotted the river way down there below. The old man, who was still learning about patience, answered over and over "That's the Fraser River, John." I must have been a slow study, or maybe the immensity and power of that great entity, moving slowly below us, struck me deep.

The trip takes me over the high Spuzzum Creek bridge. I turn off below the highway to backtrack down the old Cariboo Road, treed-in now, overlooking the hillside across on the south side of the creek. The low bridge we would have crossed in 1946 is long gone. The immediate area on both sides of Spuzzum Creek is Indian reserve land; the people are the southernmost Nlaka'pamux, a word I continue to find impossible to pronounce. I stop and wonder that 120 years ago, approximately, this seemingly unremarkable slope, looking now like just another second-growth sidehill, was a concentration of gardens and small farms, an outcome of missionary efforts to turn Indians into farmers, no matter how cramped and rocky the local ground. A roadhouse had done business nearby, and miners and settler's cabins, Chinese and European, stood by the creek mouth even before the Royal Engineers' road was completed. Long before that there were numerous pithouse homes up and down the creek. Lives were lived here. As in so many backcountry places, there were more people here in pre-contact or early contact days than now. A succession of European diseases, of course, was devastating for aboriginal peoples.

In 1858, this stretch of river was the scene of the little-known Fraser Canyon War. Miners, many of them forty-niners, recently up from the depleted gold diggings in central California and used to anarchy, and a few, perhaps, fresh from hunting Indians there for sport, were looking for gold in the lower Fraser River. Assaults of one kind or another were committed against aboriginal people; at the very least, miners presumed their natural-born right to use and take Indian land and resources. Local Nlaka'pamux resisted. The "demon booze" was a factor on both sides. The miners, some aggressive, some fearful, banded together and, employing a strategy proven useful down south, organized themselves into would-be militia units, then initiated skirmishes against local aboriginals. The natives of the canyon, no strangers to struggles over territory, and numerous, fought back with vigour; it was their home ground after all. Bodies, both Indian and white, floated down the river.

James Douglas was concerned that violence in the Fraser Canyon would lead to lawlessness and more violence, just as he was concerned about McGowan later in the year. In the summer of 1858, in order to maintain control and assist in the protection of the aboriginal inhabitants of the canyon, Douglas took it upon himself to act as if he were governor of the territory. He declared that miners along the Fraser must buy a licence before they proceeded further; he commandeered the few military personnel around to help with enforcement if necessary; and he kept in contact with London as closely as he was able. Meanwhile, the British government was moving to grant the mainland colonial status under its direct jurisdiction, with Douglas as governor, though it would be months before he learned of his formal appointment. A small detachment of Royal Engineers was sent to the new colony as quickly as it could travel, and Matthew Begbie arrived in the fall to take up his duties as judge in the new colony. Once again, the fast, decisive actions of Governor Douglas to establish British control over the goldfields were crucial in the creation of what would eventually become the province of British Columbia.

Just north of the Spuzzum Creek turnoff, I spot the small white house and grown-in garden and fruit trees, on a small bench between the railroad tracks and the river, where Annie Zetko York and her cousin, Arthur Urquhart, lived for many years. Annie died in 1991, not so long ago really; Arthur more recently. It looks as if nobody is in the old house now. The roof is mossy, trees lean in, the house looks a little darker and a little smaller. There are still signs, though, of Arthur's fine detailed maintenance all around.

Annie was somebody: a woman of considerable intelligence and knowledge, who bridged the old times and the new. Like many Indian people, Annie was of mixed ancestry. Her paternal grandfather was Cataline, the famous pack-train man, but she was raised and educated primarily by her Nlaka'pamux grandmothers. She was evidently curious and a learner, and she became very aware of who she was, where she came from, and who and what she was connected to, including the country down and especially up the river, where her aboriginal ancestors had lived "since time immemorial." Annie was spiritually minded and very devout, and like many of her generation she accepted and practised both Christianity and Nlaka'pamux beliefs. She was much influenced and felt protected by an aunt from the Nicola Valley, Josephine George, a healer. Annie seems to have gradually become chosen by her Spuzzum elders to be a carrier of the old knowings, in part

The little white house and garden below the tracks at Spuzzum where Annie York and Arthur Urquhart lived through much of the twentieth century.

because of her accessibility; the irony is that if she had been of formal Indian status, she would have been away at residential school for much of her early life.

To the best of her capacities, over a long lifetime of hard work and hardship that included two near-death experiences, Annie was faithful to the spirit of her old folks. To exemplify, when asked, she could read the many Stein River rock painting panels like a newspaper — or the way my investor friends read their monthly stock statements. She was particularly knowledgeable about plants and their uses and became an important resource for aboriginal and non-aboriginal students alike, most especially, perhaps, the renowned ethnobotanist Nancy Turner.

Fortunately for old-time knowledge, and for us all, she has been well documented. Her Stein River interpretations are the major chapter in *They Write Their Dreams on the Rock Forever* by Richard Daly and Chris Arnett, with Annie as co-author and prime source. As well, she co-authored, with anthropologist Andrea Laforet, *Spuzzum*, an amazingly detailed and thorough book that contains information about Annie and her extended family. She is a major source in many more. When she died, an essential way of knowing, though not all her knowledge, died with her.

By the old house the river is quiet.

In *Bowl of Bone: Tale of the Syuwe* by Jan-Marie Martel, a film substantially and movingly about Annie and Arthur, there is a series of frames of the river flowing quiet, dark, a picture of slow power. Judging by the light, the season is summer, and warm. This particular image is strong in my mind and connects to a parallel sense of those two old people and other elders of their era, both Indian and non-Indian, like them. At times that image of moving, dark water in bright light and warm air fills me with a nostalgia so deep I could choke. In this unattached cultural world of speed, acquisitiveness, waste and no silence, where the illusion of separation and the cult of the individual predominate, I mourn something lost and gone. I mourn for that old, slow world of knowing, where children could grow up with time to learn to think and understand, and where we were smart enough, some of us, to learn to recognize that all elders know something.

These years I drive the canyon highway fairly frequently and, coming and going, find myself saluting the old Urquhart house, half buried in the trees, as I pass. I must in some way acknowledge Annie York and her knowing, Arthur Urquhart and his support and competence and his collection of old railroad watches, and all those older-generation folks who paid attention to where they lived. Their living demanded a measure of mindfulness.

Gary Snyder, the poet and student of Zen Buddhism and west coast culture, tells of being in the back of a moving truck in the Australian outback with an aboriginal elder who was speaking very rapidly to him, non-stop, not at all the normal mode. He was telling Gary details of the landscape and dreamtime, one story after another. Gary was mystified for a while, until he realized that any traverse of the country was a "walkabout" of sorts, and the old people were bound to recite the ancient place-based stories as part of the obligation to acknowledge and name the old ways and beings. It is good manners. But, Gary's aged companion had to hurry because the vehicle was moving faster than people walk.

Having learned a few details of older times and places here and there across this land, and especially up and down this river, which is so important, so present, and where so much has occurred, ancient and modern, I feel bound instinctively to do something of the same. That is, I feel I must say a little of what I know as I pass through. To recall briefly some history and details of the place, as clearly as I am able, is to recognize and acknowledge; it is, as much as anything, a matter of gratitude and respect. It is also an acceptance and a witnessing that we are inevitably and mercifully in a context that is lively and larger than ourselves. It is a moment of selflessness, as giving thanks always is.

I come to whiteman's Spuzzum, the sapper's streets with a few house trailers on them, one or two old railroad-age buildings, and the blackened gap where the gas station and cafe used to be, with the sign that said, "Welcome to Spuzzum — Thank you for visiting Spuzzum" in pretty much the same breath.

I cross the river again. If, as you cross the high bridge past Spuzzum, you look quickly and sharply upriver, you can see the second Alexandra bridge for a moment (in use now only as a footbridge), the very metal-decked structure my family and I would have driven over on our way up to our new Interior home. The land on the west side is Indian reserve land called Titkwalus. There is a photo in the book *Spuzzum* from approximately ninety years ago that shows the original bridge and a small cluster of houses at its west end below the CPR tracks. One of them belonged to Amelia and Joe York, grandparents of Annie, who she lived with at that time. This place was a fording and fishing place before the bridge was built; gill nets for salmon are set, in season, in the back eddies below the bridge to this day.

Higher up the hillside, where the ground levels off somewhat, is an area once called Crowsnest that was cleared and farmed by Nlaka'pamux people for a few decades around the turn of the twentieth century. There was constant need for stock feed, and tillable land was limited in this stony canyon country. There had been houses up there also. It is thick second growth now, but the signs of old human activities are every-where: stumps, the old sidehill road up from the bridge, log riprap on the downhill side, a vintage

It is a moment of selflessness, as giving thanks always is.

car body, mouldering boards in the underbrush, ditches, tin cans, green bottles, Just above the bridge are lilac and apple trees, and irises planted by Amelia York, still blooming in spring as they were several years ago when Peter Stein, one of my walking friends, and I were there prowling around, smelling out history.

Past old wire-fenced cemeteries, side by side, each marked with a large, wooden cross, stark white against second-growth firs; past old Alexandra Lodge, still standing semi-derelict in the dogwoods, one-time purveyor of the best hamburgers on the Cariboo high-way; and past a gravel cut, there is a worn-in pack trail branching off acutely uphill. This was Alexander Anderson's Hudson's Bay Trail, an attempt to bypass some of the steeper parts of the canyon, namely Black Canyon and Nicaragua Slide approaching Hells Gate. I've walked up it a few hundred yards to get the feel. I would like to get up on top before too long, while my legs still work. There was some sort of roadhouse, Lake House, up there beside the chain of little lakes along the ridge, but all signs of it have now disap-peared, so they say. Apparently deep snow was a problem.

At Hells Gate I recall my own revisitation and experience of that place and the deep, fast-moving river running through it. I had been working for months on the rigging or running dozer boat at Mahatta River, a lonely logging camp at the mouth of Quatsino Sound on northwestern Vancouver Island. It was the summer of 1965, the weather was hot, the forests were tinder dry, and we shut down for fire season. When I got down to Vancouver, acting on impulse, I rented a car and drove up the Fraser as far as Hells Gate, a place I had not seen since I was a young boy. The canyon was in shadow when I reached it, but its steep stone walls were very warm, and I could feel the accumulated dry heat of the day radiating from it, such a different possibility from up coast. I was struck by the dark gorge and the way the water, also dark, slowed below the narrows, cool, flat, silent and deeply powerful. The river stands on its side to push through at Hells Gate, and I was caught by it. That was the first step in my return to the BC Interior.

Before Boston Bar, I turn off on a trunk logging road east up the Anderson River. In Cole Harris's *The Resettling of British Columbia* there is a chapter about the Fraser Canyon that includes a most interesting old map of Boston Bar, circa 1913. The map shows various surveyed lots, Indian reserve lands and one reserve lot in particular on the Nicola Trail, away from the Fraser, up the Anderson River. Several points are intriguing here: the Anderson canyon is very deep and steep; the trail obviously goes to what is now Merritt and old Fort Kamloops, linking up with the Hudson's Bay Brigade Trail; and the reserve parcel is in a south-facing basin, seemingly in the middle of nowhere, but for the fact that few bits of surveyed Indian land were "nowhere." When land was allotted there would have been either a food-gathering or food-growing function, or an Indian person would have been living on it at the time.

The Interior plateau breathes and waits.

The father of Gordon Antoine of Coldwater (who was my boss when I worked for the Union of BC Indian Chiefs in 1973) had been a horse packer over that same trail in the old days. I had met the elder Mr. Antoine, a stalwart old man, standing before me as if rooted, looking like he was made out of rope. I had to go check out for myself where that old trail went.

After winding steeply up and around a big bend above the canyon, I came eventually to that south-facing cove beyond it, the landscape grown in as usual. But where that Indian land was designated there is a small house, clearly lived in, with a hummingbird feeder hanging, sprinklers sprinkling on a small hay flat, and no stumps in the underbrush around. This must have been a stopping place for workhorses in a hay-poor country in the old trail days. Horses would have been tired from the climb up from the Fraser or down from the backcountry. There were Indian entrepreneurs even in early times; maybe one lived here. I crawl back down to the highway and remind myself not to speed on that long straight stretch, a perfect place for traffic tickets, into Boston Bar.

A side trip over the river again to North Bend takes me up Westside Road past a few farms, an Indian cemetery near Speyum Creek and the old railroad stop at Keefers. After some miles I stumble onto the most amazing little graveyard in dark, overgrown woods, with headstones and details of several pioneers of obviously Hawaiian background. There is a Kanaka Bar farther up the Fraser, a Kanaka Creek near Maple Ridge in the Lower Mainland, and some aboriginal people from Lytton claim to have Hawaiian lineage. There are numerous reminders of Hawaiian settlers — former Hudson's Bay employees — and their descendants all around this province, on the Coast and in the Interior both, and a few of those "Kanaka" descendants are laid to rest right here. Who would have known? The sun's rays are slanting through the timber, lighting one or two headstones. Time is late. I return to Boston Bar and resume my way north.

Coming up the canyon, this crack between the Coast Range and the Cascade Mountains, is like a slow entering into the body of British Columbia. You don't come over those sheer, thickly firred mountains; you come through them. The river's rock-wall flanks on either side close in behind you. The Interior plateau breathes and waits.

I am playing road music, Norwegian folk music, on my tape deck, wild stuff from those high-latitude fjords, valleys, and mountains — bright in summer, I imagine, and drizzled, drenched and dark the rest of the year. The music is appropriately northern; that is to say, melancholy at times, deep, full of heart, with bent and piercing minor tones, the kind of music that comes out from between the black and white keys like blue notes, stops us and makes us listen and hear. It is enough, sometimes, to just about drive me off the road. Many, maybe most, of us northerners are at least a little melancholic; it's part of dealing with the dark, the cold, the duration, those long winter nights when the spirits come creeping in close to listen and connect. Street-lit city life blurs our experience of the depth and meaning and precariousness of the dark season.

I think of the loneliness of immigrants in an unknown land, the loneliness of both my English grandmothers, for example, separated for virtually all time from their families, their homeland, the places where they played. My paternal grandmother, Nina, died young, though not healthy, in childbirth. In her last photos she looked thin and wan and unhappy. The child, who would have been my uncle, died also, nameless and unacknowledged on his mother's gravestone in Ross Bay Cemetery in Victoria, where my mother and father are buried now. Nina's last words, as passed down, were "and look after little Pat"; that would be her living son, my father, not yet five years old when his mother died. It was 1918, and her husband, my grandfather, an officer in the Canadian army, was still overseas. Violet, my maternal grandmother, who was chronically dissatisfied, irritable, dis-eased and a touch fey, lived a long life and died a demented and fancy-filled death, united in her imaginings with her old-country family and relations at last. The sins — in other words, the separations — of the grandmothers (and grandfathers) are visited on their sons and daughters, and there is a tinge, a layer of that old upheaval, disconnection and heartache in many of their descendants today. It is like a narrow stratum in the undershorings of our being.

And day and night, spring, summer, fall, winter, no matter what we are doing, no matter what our great endeavours, our tricks, our tiniest solipsisms, our acts of true love, the river flows. There is something in its constancy, its power, that draws our deepest attention, holds it, preserves the image of it in our being and shapes our seeing forever.

Some years back, I was driving down the canyon, returning from one of my summer trips to the dry Interior with my very aged parents, an annual treat for them and for me. We stopped to rest at a long curve in the highway overlooking the river, below Boston Bar. It was late afternoon and the river was flowing away from us to the south in that inexorable manner, looking like molten silver mud. I was in the throes of a fairly major personal crisis at the time and feeling desperate; the journey to the bird swamps, meadows and pine woods of the Chilcotin with Mum and Dad, always their inimitable elderly selves, was a welcome moratorium. I stood at the edge of the highway, looking down the canyon, the mighty walls and slopes of the mountains as a backdrop, and watched that flowing river far below. I was feeling like an open wound, but never so sore or desperate as to lose my basic faith nor my capacity for prayer. I simply asked the river for guidance and waited. The river flowed away, and a minute or two later the words came like a balm: "The course of least resistance to the Sea."

I acknowledge that place at the bend in the river each time I drive by. Occasionally I have stopped the car and walked to the edge to gaze at the big silver river flowing away down below and murmur a few words of gratitude. Once I left a small twist of tobacco tied to a low fir sapling as a statement of acceptance and an act of respect.

Just short of the Canyon Cafe and truck stop, where I stop for coffee — an old routine — is the first ponderosa pine. It is a large, lone, sturdy tree standing, with its characteristic eroded red bark and its blurry bundles of blue-green needles, on a rocky hillside back off the road, looking truly "ponderosa." A little farther up the flat there is a whole grove of them, young and green and healthy still, on both sides of the highway. I will see the first big sagebrush soon as I approach Lytton, and king birds, larks and magpies on wires and fence posts, and the occasional coyote. Now I know I am in the Interior.

THREE

In the Details

*The more of a treasury of detail you have in your
mind to draw from, the stronger the work will be.*

PETER MATTHIESSEN

My long-time friend Tom Hueston and I reached Empire Valley
by mid-afternoon on May 7. Winter had been mild but late leaving, and the snow line was low. Weather was bright and cold for
May, and the place was full of mule deer in groups of three to
thirty or more, still wintering, waiting to make their move west
soon, up into the high places. Where there were that many deer,
there would be cougars, seldom seen, of course, at least by me. I
considered myself fortunate to be able to identify a bit of twisty
cat scat I spotted on the roadside.

We were at 2500 feet above sea level, approximately 1500 feet
above the bottom of the Fraser Canyon. The first leaves on the
poplars, willows and saskatoon bushes around Empire were just
coming out with that wonderful delicacy and vulnerability of
early spring, and grass was greening. The Chilcotin plateau levels off at something like 3500 feet, however, and this would be
the last of spring showing until we reached Tatlayoko Valley and
West Branch, our eventual destination, sloping down through the
Coast Range 120 miles or more west of us.

The air was clean and clear across the river to Canoe Creek and
down the dry benches and gullies to China Gulch, Crows Bar,
Lone Cabin Creek and some of the loneliest bunchgrass land
in BC. The contours of the bare land were knife-edge sharp, as

FACING PAGE:
*Empire Valley
Ranch horses
on Clyde
Mountain.*

31

only chill air can convey, and inviting. Tom and I were keen to do some prowling around.

We drove up Iron Gate Road to get a peek at the open bighorn benches and escape breaks up along Churn Creek. As well, I wanted to see if we might get some sense of the whereabouts of Stranger Wycott's place, an old homestead I'd been hearing about, which was presumably in the vicinity of Wycott Flats. References to Wycott have always intrigued me, not only because of his ear-catching nickname, but also because, even though his name comes up fairly often, the location of his place generally seems to be a mystery. Where is it exactly? What does it look like? I'd heard also that wild sheep, Californias, wintered there and around other flats up and down Churn.

I wanted to show Tom the Chinese ditch, a laboriously engineered, five-mile watercourse dug in the 1880s from Koster Lake, once known as China Lake, off the dam they built at the outlet, down to separate placer gold mining operations at Onion and Fisheries bars on the Fraser River. The immensity and isolation of that task invariably moves me. I can imagine, too easily, all those tree roots and boulders to dig through or around, all that food gathering in a strange land, all those hostile neighbours, white and Indian.

Not far from where the ditch splits to go around either side of Airport Mountain, we found the original Brown Ranch homestead, on land pre-empted by John Brown in 1893, still standing firm in the poplar thickets above BC Lake. James Nathaniel Jerome Brown, another of several Brown brothers, published a book of poems in the Robert Service mode later in his life after he had moved to North Vancouver. As a serious would-be sculler, James is remembered also for packing a rowing shell more than ninety miles over the trails from the railhead, at Ashcroft in likelihood, into Empire Valley. He made the trip with the help of a friend, and, I sincerely hope, at least one horse, and practised his rowing up and down Brown Lake in Empire Valley. James and John's father, Samuel Leander Charles Brown, built and owned a grist mill believed to be powered by the original Chinese water wheel at the creek outlet at China Lake, in anticipation of settlement around Empire Valley. He used the same burr stones that had been in the mill he and his former partner, Le Comte Gaspar

Looking south down the Fraser River from Clyde Mountain in spring. China Gulch is across the river, Crows Bar is around the corner on the same side, and Lone Cabin Creek is in the far background on the west, (right) side of the river. The Empire Valley Ranch headquarters are out of sight to the right up Grinder Creek.

de Versepeuche (Isadore Gaspard), had owned and run across the river at Dog Creek in earlier years.

Tom, ever the artisan, studied the close-fitting log corners of the Brown house, and we tried to find the refuse dump. Old garbage is always a curiosity. We searched out and found the grown-in wagon route through the willow bottoms to the north. It led to lot 311, pre-empted a few years later by one of the several Bishops who settled the area to take advantage of the water supply flowing down the ditch, which was still functioning at that time. I will need to get down there sometime and hunt around for signs of the old home-stead: the barn and house sites, the root cellar, alien barnyard plants, tin cans, old bottles, strange intuitions, wild ideas. Logic tells me it would be close to that ditch somewhere. I've already spotted hundred-year-old fence lines: rotted, sawed and notched log ends and a bit of twisted wire here, a line of trees and a pile of rocks there. The old cart track running down into thick standing timber alongside the north end of the ditch looks like it should lead right to it.

On Hairy Fish Lake the first goldeneyes and buffleheads were bobbing their heads up and down, the males one-pointed, aggressive; the females seemingly oblivious. The smallest ponds and cat-tail holes were called lakes in this sagebrush land. Here, if a lake dried up for a portion of the year, it would be named, naturally, Dry Lake. That starlit evening, we heard geese flying in the clear, still air above us, honking advice and reassurance to each other, the old ones leading the young ones as they passed over on their way north.

We drove back down into Empire Valley, and I showed Tom a little rectangular flat below Brown Lake by Koster Creek where the long-gone and short-lived schoolhouse had been, just a few weeds and wood bits indicating that it was ever there. Across the road, at the site of what was the Boyle homestead a century ago, there was nothing left but congregated rock and rearranged dirt; near-invisible fence lines separating grass types, wild on one side, domestic on the other; and a little cluster of small squares on the 1:50,000 topographic map for the area. I could hear no children's voices, not even a murmur.

On the other side of the track, near the edge of the upper end of Bishop's fields above Brown Lake, is a leaning, two-storey log barn, a few rhubarb plants in a row nearby and a lone apple tree, the last remnants of that family's once famously flourishing garden. Beyond and above the barn on a sidehill stands Sam Kenworthy's house, a broken-backed shell now. Kenworthy was half-owner, with the Kosters, of Empire Valley Ranch, having inherited his share from his mother. The young man's father had not come back from the Great War, and his mother could not maintain the place on her own. Her neighbours had been interfering with the irrigation water, trying to squeeze her out. Young Kenworthy built his home-to-be with high hopes and great care for his new bride, fresh out from England. Despite his efforts to provide such civilized touches as a built-in kitchen sink with outside drain, wallpaper, painted door and window frames (white and robin's egg blue), the requisite lilac bushes, two double-doored root cellars, and a glorious view to the southeast down Koster Creek, the separation and isolation were too much for her. Their first winter, an especially cold one, was hard, and he was apparently too "toffee-nosed," too conscious of his class, to get along with rough-edged Chilcotin cowboys. They moved to Vancouver Island to raise mushrooms not long after.

Nowadays, balsamroot sunflowers grow across the slope behind the Kenworthy house, blooming in great, golden profusion in early June, protected from grazing stock by the stout, barbed wire fence all around the place.

I looked down the valley to Graveyard Field and wondered what the story of the graveyard was and where the graves might have been. That field could have been the site of the camp of Shuswap (Secwepemc) people, members of the Empire Valley (Tcexwe'pkamux) band, who were massacred by Lillooets (St'at'imc) from downriver about 1825 as stated in an appendix to the management plan for the Churn Creek Protected Area. The primary source of First Nations information for the appendix was James Teit, the

A coyote had left a recent calling card inside the barn.

famous ethnographer from Spences Bridge, who in 1909, with the backing of the great anthropologist Franz Boas, published a comprehensive study of Shuswap life in early contact times. Lower Empire Valley, where Koster and Grinder creeks come together, midway between the watered and timbered uplands and the hot, sagebrush river bottom, is the usual kind of location for a winter village. Or were the burials casualties of the smallpox epidemic in the early 1860s? The graves may even have been more recent still, settlers perhaps.

At the calving barn, a large, open-ended structure surrounded by a labyrinth of aging corrals and holding pens, now designated as a camping area, we parked and Tom cooked up a late stir-fry. A coyote had left a recent calling card inside the barn, in the middle of a piece of cardboard right next to the picnic table. My partner Marne and I had used that very cardboard the previous summer, stepping onto it from a makeshift, and brief, cold-water shower bath, a respite from the heat that day.

After supper, we went down the valley toward the ranch headquarters and counted mule deer in the lower hayfields. We got up over three hundred before the evening became chilly and it was time to leave. Tom said it was like they raised mulies here, not

Upper Empire Valley, looking down Koster Creek toward Brown Lake, circa 1912.
The Bishop family's ranch and garden were located partway down the valley at the
near end of the lake.

cattle. These mid-West Fraser River slopes are some of the greatest and wildest wintering grounds for deer in the province. There
are California bighorns in this area also, but we saw none. The
·ewes must have already been on their various lambing grounds
in the breaks and coulees above the river. As dusk was falling, we
drove back to the calving barn and turned in.

I woke at first light with the fresh and sharply invigorating news
that my old down sleeping bag, having served me well for decades
but losing two feathers per sleep as usual, was just not cutting it
anymore. I warmed myself by getting up and walking, ostensi-
bly looking for early bird migrants, along the willow bottom and
meadow up from the barn. The ground was frost white, and the
few small, nameless, grey-brown birds I spotted were all fluffed
up and silent, facing east, anticipating the sun's imminent appear-
ance from behind the ridge. Cooking breakfast, my hands hurt
from the chill. I hadn't thought to bring gloves — it was May, after

all. Mysteriously, the coyote poop had disappeared; pack rats, I suppose. I had heard their rustlings and scuttlings in the previous night's dark.

On our way north through Gang Ranch after packing up, we passed approximately thirty scimitar-beaked curlews in an irrigated field below us. The field was striped in bands of frozen white from the sprinklers, and the birds were feeding intently, some in sharp, slightly surreal profile against the ice, on worms and insects brought to the surface by the previous day's water, I presume. A little farther, by one of the barns, a big old lilac bush was jammed with bickering yellow-headed blackbirds, a barrage of yellow-headed squalling. I had seen and heard the same thing at the same place the spring before. That particular bush must be one of their regular migration stops. With them, for the company and safety no doubt, was a single sparrow-sized redpoll, headed for the Arctic.

Tom and I drove by rolling hay fields and over open, sloped grassland in behind the Gang, as it is known, ascended into snow, road muck and Douglas fir for a few miles, later dropping to cross Little Gaspard Creek by Augustine's and William's meadows. The creek was in full spring freshet. Just before the meadows, an old logging road branches southeast, and I was thinking that somewhere off that road, on the far side of the rise, there should be some kind of walkable access down into the grassy benchlands and steep breaks and gullies along mid-Churn Creek. One of those benches would be Wycott Flats, and somewhere down there William "Stranger" Wycott had settled, fenced and built and made a living for his family. I was developing an urge to drive over there sometime to search for a route to his old house and barn, intact and still standing some say. An image of those buildings was beginning to form in my mind.

Eventually we came down into Farwell Canyon, with its once-farmed flats and a pair of well-built, abandoned ranch homes on the near, upriver side of the Chilcotin River bridge. That former ranch is referred to as the Pothole. There were cottonwoods along the river, a creek gully and red willow jungle behind the houses, and an Indian salmon camp buried in the willows by the creek. There had been a rock carving, an aboriginal petroglyph, on a slabbed beach rock by the cottonwoods, thought by some in older

times to have marked the boundary between the downriver Shus-
waps and the Chilcotins (Tsilhqot'in) to the west. But in my time
some collector stole it, and its theft has cost us all the privilege of
seeing it.

There are stories here, layers of them.

According to the ethnographer James Teit, there were two old
Shuswap winter villages below the canyon, one on either side,
each a complex of kekuli holes, or pithouses. A third village site
was located farther downriver where the Chilcotin joins the Fraser. The area is known as the Junction, and these people were
known as the Canyon Shuswap. They and the Empire Valley band
were the only Shuswap-speaking people living west of the Fraser
in pre-contact times, and the nineteenth-century diseases were
especially devastating to them, probably in part because they lived
on a major throughway. The great Fraser always was and always
will be a conduit for traffic of all kinds, birds, people and germs,
passing north and south. Members of two Shuswap families continued to live west of the river until modern times, but the rest of
the remnants of the Canyon band retreated across to Alkali Lake,
Dog Creek and Canoe Creek. Chilcotins from the west moved
in to take over most of the territory. Jimmy Rosette, a descendant
of one of those two remaining Shuswap families and a long-time
Gang Ranch cowboy, is supposed to have told the ranch owner of
the time, "You can't fire me, we've always lived here."

There is an area above a broad bench on the north side of the
Chilcotin River where I sometimes see California bighorns at the
tops of the draws, mostly ewes and young chewing their cuds and
checking out intruders, or on the slopes and benches feeding.
Once, at dusk, I saw a big ram walking up a point. All I could see
of him in the gloom was his pale rear end, but I knew it was a ram
from his size and, especially, the heavy insouciance of his walk.

From that same spot you can look downriver to a wide, hot-look-
ing, grassy bench on the south side. Exactly on the eroded gravel
edge above the river, one corner overhanging, looking unusually
stark, is a sun-blackened log shack waiting to fall in. In her book
Chilcotin: Preserving Pioneer Memories, Veera Bonner, one of the
Witte Sisters, third-generation descendants of an original Chilcotin rancher, mentions that that cabin belonged to Joe Dan Smith,
who worked at the ranch and road house at Riske Creek known

as the Becher Place during the 1890s. Dan Smith was black so, predictably, he was known as Nigger Dan. He was well liked but was sent to jail on a shooting charge and was not seen again. In his memoir *Cariboo Cowboy*, Harry Marriott talks of riding north to visit a friend, Jim Ragan, whose residence was nestled in a high draw overlooking the bridge, and meeting one of Smith's young, curly headed descendants among a family camped below.

At the back of Smith's bench, where the steep canyon walls begin, there is a trail, well dug-in but now little-used, that is quite visible from across the river. It angles sharply and firmly straight up the hillside. Years ago I wondered about its function. Later I realized that it came from those old village sites beyond the canyon, probably at a crossing-place in the river. It is another of those ghostly 1000- to 10,000-year-old trails that crisscross this country. Old trails live, and the way to keep a trail alive, I'm told, is to walk on it.

In her book, Veera Bonner describes an unusual place farther downriver from Smith's bench, on the west side, that was built by an Englishman named Charles Manual Castellain. The love of his life was a Chilcotin girl, and he built a cellar-like underground home, much like an old-time pithouse, for her. I have seen the old place, still intact in its forgotten corner by Castellain Springs. In

Sam Kenworthy's house, built in the 1930s, inhabited by the Kenworthys one season and used little since. Empire Valley and Brown Lake are in the background.

fact, I crawled into it. It seemed like a close and damp place to me. The fair Mrs. Castellain died of tuberculosis, like so many then. Castellain left the country and died in Vancouver not long after, heartbroken, so the story goes.

Two British mudpups, Mike Farwell and Gerald Blenkinsop, built the two solid homes in the Pothole and farmed and raised stock, mainly horses, on the low, fertile land there about ninety years ago. The place is powerful and picturesque at its bend in the river, the windblown clay walls and hoodoos on the opposite bank, the high sand dunes to the northeast. But it is a hot, dry place in summer and was isolated in winter back then.

The partners had bought the land from Louis Vedan, whose mother, Mary, was aboriginal and whose father was likely a gold seeker from France. Veera states that he was known to march to his own drummer, irrespective of clocks, the season and contact with the world at large. After separating from his second wife and living for a while at Sugarcane Jack's on the road west from Farwell, Vedan eventually took up land again in the Whitewater country southwest of Big Creek. The cabin and barn he built over there are still in use. In summer he ran his cows on the grassy, southwest-facing slopes and flats along Elkin Creek by Vedan Lake near far-off Nemaiah Valley. From his Whitewater place, the trail to

The calving barn and corrals were built and previously used by the Empire Valley Ranch.

Elkin Creek, following the grazing meadows along the swamps and drainages, would have been tortuous and rough going. This was a man undaunted by distance, hardship and, apparently, by time. It was Louis Vedan who was so taken by Gerald Blenkinsop's new wife, Queenie, and her "bright presence," as Veera puts it, that he bought her a grand piano as a wedding gift — a Gerhard Heintzman, no less. Queenie Blenkinsop later lived in Williams Lake for many years and died in 1996 at the age of 106. I can so easily imagine and hear the musical beauty and good English cheer rising, like plumes of smoke, out of the secluded depths of the Pothole back in those days. The settlers, fresh from the Old Country and new to this raw place, would be busy keeping the spirits and goblins and other denizens of the cold dark at bay with laughter and the trappings and artifacts of civilization.

At dusk, I saw a big ram walking up a point.

The seemingly endless winters must have been so hard for the first settlers to the cold Interior and up the wet West Coast, and at times for aboriginal people. Folks in the Chilcotin still refer to getting through the dark season as "wintering," or conversely they might say, "He didn't winter this year"; in other words, "He died." When wild game was scarce and the dried salmon ran out, life became very difficult, and there are stories of starvation in the not necessarily good old pre- and early-contact days. The great British Columbia artist Bill Reid, in *Out of the Silence*, remarked that the winter ceremonies on the coast and up the Fraser River occurred, in part, because of the people's need, "no matter how long their lineages," to reassure themselves of their greatness against the "long black winters."

It is the first settlers like Louis Vedan, though, who most thoroughly capture my attention, because he, like a number of other early settlers of the East Chilcotin, was thought to have been born, or at least raised, in Empire Valley a long time ago. Lillian Bambrick (nee Haines), who married Charlie Bambrick, first settler at Big Creek, is another from Empire Valley. The fact of the existence of these people piques my sense of the misty beginnings of that place. I have seen the location of the first Bambrick cabin, a rectangle of low foundation stones that is hard to spot in tall grass at Hutch Meadow on Minton Creek below Fletcher Lake. Veera

Bonner had to show me twice how to find it. Standing there, it is mildly unsettling to recall that Walter, oldest of Lillian and Charlie Bambrick's children, and one-time cow-boss at Chilco Ranch, came into this world within the dimensions of that modest little cabin site. Judging by when these people settled their land or worked their jobs, their mothers were giving birth down there at Empire Valley in the 1870s or even earlier, before the place received its modern name.

We know that, in the wake of the Cariboo goldrush, William Wycott twice pre-empted land in the area — first down on the Fraser River, on the east side, in 1868, and again in 1884, up Churn Creek somewhere around what is now Wycott Flats. I'd been hearing and reading bits and pieces about this Stranger Wycott character, hints that he was unusual, suggestions that he had become "bushed" and that his livestock was overgrown and completely untamed. In particular, there are curious stories about him from his declining years when he seems to have become connected with the Gang Ranch. I wondered who he was, who was with him, who and where his offspring were and just exactly where his rather mysterious homestead might be. More generally known were the Harper Brothers, up from Virginia and the American Civil War, who founded the Gang Ranch on Gaspard Creek early on, although opinions differ widely as to exactly when. The standard knowledge is that the ranch was started as early as 1862, but historian Don Logan, associated with the Clinton Museum, has researched the question and states that it was pre-empted by them in 1883. There is no recorded acknowledgement of any farm buildings there before that time. Considerably more people lived in the Empire Valley, Gang Ranch and lower Chilcotin River region a century ago than now, and the area must have been lively. Resonances of that liveliness, as well as the antiquity itself, have fascinated me for years.

As the sharp edges and details of historic knowledge fade over time, some characters and places, mountains, trails, old buildings, even certain animals, wild or domestic, take on the larger-than-ordinary-life qualities of myth. To experience myth in a lively way requires a certain whole-minded attitude, an ability to be in the present tense and stay in it, a willingness to not so much think about as absorb our experiences and awareness for a while, and a

capacity to embrace the numinous. Energetic walking alone in wild and silent places over time and asking the kinds of questions that places, especially wild places, can generate may help to put you there. Feeling gratitude and saying "Thank you" helps us to be modest, as does a sense of awe. We might call such a non-judgmental, open state of mind "myth mind." Such myth-mindedness complements literal awareness very easily and usefully, and it can expand our experience and understanding of meaning in our powerful and dynamic world.

Gang Ranch headquarters on lower Gaspard Creek.

Even small details of a half-known past can be historically and mythically evocative, and they draw me inexorably: a shard of blue-willow-pattern crockery in a moss bank; square nails protruding from a living pine-tree fence post; a woman's shoe half buried in yard dirt; a decaying piece of harness hung from a branch and hanging there still; a single green bottle on a stump.

For an outsider, I have learned a surprising amount after forty years of looking. But as I say to people, I haven't lived here in winter, ever, and I'm not likely to, so what do I really know? My knowledge might be 300 miles wide, but it's about an inch deep. We are just beginning here.

I began by simply trying to get a sense of the Chilcotin country, vast as it is, one-third the size of France, populated by just a

few thousand people, the closest we come to frontier's edge in
the southern half of BC. I found myself asking, with an increas-
ing obsession: What is around that corner? What's up the next
valley? Where does that creek run from here? What would you
see from the top of that ridge if you could just get up on to it?
There must be an old route down this drainage somewhere. Who
would have walked through here in the old days? Where do those
tracks come from? Where do they go? Does this trail go right the
way back to the ice age? Who lived in that old shack out there all
alone? Why is there a long-used campsite here?
And why here and not there? Where does that
treed-in wagon road go? Where does this ditch
come from? Who dug it? And why are there old
sap-extruded, chopped (not sawed) stumps in
this particular place in the middle of absolute
nowhere? What is this feeling tugging at my innards so persis-
tently? How is it that that shaft of sunlight lights up that precise,
old blaze mark so brightly? Why do I hear certain whispered rustlings in the golden, long-shadowed, late afternoon when the chill
is creeping up the draw? Why have I felt the need so enduringly to
try to put my questioning experience into words?

'True knowing
begins with feet'

Over the years, I have come up to this country often, usually
with others: my parents when they were alive, my brothers, my old
university friend and longtime backcountry partner Don Brooks,
my Victoria friends and hiking pals and of course my dear Marne.
But now, increasingly, I choose, sometimes, to walk these places
by myself. Peter Stein from Victoria, a determined hiker and an
enthusiast for the American Southwest, gave me a book recently
by Reg Saner, a man who travels and walks alone and who has
walked certain portions of that dry southern desert country for
years. In that book, *Reaching Keet Seel*, he makes a statement I
cannot resist: "For me, always, true knowing begins with feet."

I have done some kind of desultory but persistent practice nearly
all my adult life. By "practice" I mean regular periods of quiet
mental focusing, either sitting or walking, as performed by medi-
tators around the world, aimed at increased mindfulness, balance,
effectiveness and pleasure. Walking has become one of my main
practices: walking, meditating, reflecting and attempting to pay
attention. One of my older and deeper instincts is that, as recent

settlers on this North American piece of ground (I am third generation, myself), our life's practice should incorporate, in some way, attempts to become more conscious, and ultimately more respectful, of this living ground we walk on, these places where we put our feet. That ground becomes a kind of absolute. The logic is simple: we are not here, materially or any other way that I can imagine, without it. It is us. We are it. This great presence beneath us and around us that we take so completely for granted is you and me. That is our body that we lay bare.

So it seems most pertinent to ask if we know this place where we stand and breathe and live, where, in our ignorance, we do constant sacrilege and damage, to it and to its myriad lively constituents. Do we know the lay of this land where we are: the watershed, the four directions, the places where wild animals winter and spring bulbs flower, where the weather comes from, where the birds go? Is there any knowing more fundamental than this?

These years I am walking through these Chilcotin places, some over and over, and everywhere, particularly in the eastern parts close to the Fraser — Empire Valley, Gang Ranch country, Farwell Canyon — and along the valleys in the outlying mountains — Tatlayoko, West Branch, Nemaiah — I am seeing signs. The

Farwell Canyon, on the lower Chilcotin River, and the long-abandoned Farwell ranch house.

signs become presences: rock pickings at the edge of a meadow, rotted rail fencing, century-old plough furrows on a grassy flat, a hay rick in scrub birch, barn weeds but no barn, a lilac bush but no house, the lone rhubarb clump in a field, root-cellar holes, the pile of rusty cans by a creek outlet, the Model T car body buried in brush, a sprung trap hanging from a tree branch, tether stakes alkali-white in the swamps, the wild horse round corral at timber line, a faded board with a man's name on it nailed to a tree, another trail to nowhere. And farther back: kekuli pithouse holes, stone-lined storage pits, ancient tracks leading to the river's edge, the flintlock trade musket leaning against a tree, a rock painting, a pecked boulder, the cross and grave mounds and enclosing fences on an open point, the smell of smoke, voices barely heard. And back to the beginning: a mouse-gnawed elk antler; big bear tracks, claws long; salmon vertebrae on creek gravel; white bones. I hear a single, shrill, small-animal sound in the dark of night, a coyote's yip and howl, sandhill cranes talking, slow-winging overhead, wind passing through dry grasses.

The signs are not necessarily seen, only felt: trails grown in and lost, cabins not reached, the prospector's shaft somewhere on the hillside, the dead drifter on his back in a cabin, empty eyeholes to the sky, blood poison by the lake, a pile of rocks, disease, the Indians' bewilderment, the casual shift to whiteman's land, the blackened homestead, ranchboss-burnt, thoughtless appropriation, double talk, a broken promise, brass shell casings, wild horse bones in the dark, wet earth of a high meadow, a baby's skull, intimations of murders.

My old friend Don Brooks, scientist and long-time fellow student and backpacker from the University of BC, said to me a while back: "It's not so much the answers as the questions we ask." In the process of fashioning a question, we frame answers. Questions must rise and rise again: How did we get all that way to here? Who came before us? Who followed? How did we survive? Did we suffer? How? Was there awe? Was there joy? Did we survive with gratitude and grace? Did animals inform us and help us see? When we killed them, did we do so with respect? Or were we just killing? What have we become?

Inevitably there was struggle, days and years of work, accomplishment, pride and failure, loneliness and cold and dark and dirt. There was the smell of tired bodies and excrement and rotted meat. And late-spring roses, hay curing and our lover's sweet breath. There were heart-tearing memories: babies lost and gone; mothers' drawn faces; thin-armed children; young men raging; tired-bodied, time-worn, flat-eyed elders; bloated dead animals with legs stuck stiff at strange angles; cracked earth; dry creeks and dry wells at summer's end; and the old peoples' old dry ways of telling. Do we know the accounts of their lives upon this earth? Are there heroes? Have these stories become our stories?

The spirit of such places is palpable.

In most places I travel these years, wild, having retreated, returns in mainly small and hidden ways: at Empire Valley, Churn Creek, Farwell Canyon; in the deep crack in the landscape that is Lone Cabin Creek. At Tatlayoko Lake and West Branch, close to the heart of the shining mountains and on the high slopes of Ts'yl-os over Nemaiah, wild never left. The spirit of such places is palpable. Wild predators come out of the valleys and off the slopes onto the plateau, grizzly bears on their rounds, cougar after deer, and wolves on the increase here and elsewhere across the Chilcotin. Coyotes were always here.

There seems to me to be wild at the heart of all things: in valleys little visited, in earth, in growing trees, in old places once lived in, in our backyards, in our bodies, in our unconscious minds, in all decay. Wild is that which is beyond our control. The more I watch and see, the more I conclude that, despite all our efforts at domestication of the outer world and of ourselves, we remain as much wild as tame. The line between culture and chaos is paper-thin. Read the news. Watch TV. Look at the creases on your hands and on your neck and face; listen to your heart. Even our efforts to define the word "wild" are tricky, as if the word itself resists taming. We define "wild" by what it is not, but what at its core is it? Is not a mountain, a volcano, an ocean or a robin inherently wild? Must we conclude that the essence of all existence is wild? Is there anything in our world that is not ultimately beyond our means to control? Death feels like wild to me.

"The trees are closing in, you know," my mother in the intermediate care home would say in her dementia-driven way several times a day. I could never disagree.

It was warm, momentarily, down in the Farwell Pothole as the sun climbed above the canyon walls. Tom and I stopped to enjoy it, to be in it. There were great numbers of white-crowned sparrows and yellow-rumped warblers and the first, miraculous bluebirds. This is why we come here, to see and hear and feel the first rush of spring up the Fraser throughway, the great connector to the northern hinterland.

One time, Peter Stein and I saw a lazuli bunting sitting on a willow branch, the blue light flashing off him like neon. It was in the bowels of lower Farwell Canyon by the old bridge footings above the Chilcotin River, a place steep, narrow, dark and slightly dangerous, below where salmon are dipped in summer. I have a piece of lapis lazuli from Afghanistan at home; it is deeply, richly blue, but nowhere near as brilliant as that bunting on that bright June morning. Peter and I were on an errant search for red ochre rock paintings. There were none in the canyon, but there were some on a rock face not far away, finely drawn. It took me several attempts over time to find them. I'd been looking in the wrong kinds of places. An American anthropologist, Michael Harner, pursuer of shamanic lore and experience, told me once that rock art is usually, maybe always, found near apertures of one kind or another in the surface of the earth, shaman's tunnels to the underworld, places of power and riches, very dangerous. Old-time heroes in the myth world, facing fierce, toothed animals, risking dismemberment, travelled down these subterranean passages in search of power for their people.

I had heard vague intimations of another set of paintings upstream beyond the Pothole. Peter and I searched but found nothing.

Above Tom and me, our faces to the sky, four sandhill cranes circled and circled, trying to catch the beginnings of a thermal rising off the sun-facing, north slopes of the canyon above. The one, high and light above the others, was honking those unmistakable, rhythmic, stretched-wire calls of encouragement to its friends, clearly audible above the low roar of the river.

FOUR

Creases in a Bear Cub's Paw

The Bear Wife moves up the coast.

Where blackberry brambles
ramble in the burns.

'THE WAY WEST, UNDERGROUND'

GARY SNYDER

Days later, down at Tatlayoko Lake, after we had driven up through some of that Stranger Wycott/Gang Ranch backcountry, crossed a large portion of the Chilcotin Plateau and seen some of the wild lands along the high mountain edges to the south and west, Tom decided he would fish the outlet of the upper Homathko River where it enters the lake. He had been eyeing the dark, rich lagoon off the river behind the spit at the head of Tatlayoko since we first arrived.

We had explored West Branch, been over to Big Eagle Lake and Chilko Lake, and driven down Tatlayoko to Bracewell's isolated wilderness resort at the end of the road up in Cheshi Pass. On the drive down the lake we spotted several happy black bears fresh out of hibernation and availing themselves of the first foraging along the exposed roadsides. Even though it was mid May, winter was still a presence; most of the valley was unremittingly grey-brown, and the high country was heavy snow. Poplars were just turning that early blurry grey green and only the lower-elevation, south-facing, open ground was showing spring. Here along the lake and over at West Branch, where grass and dandelions were sprouting green, deer had congregated from miles around. We have pictures taken from the cabin window, early one morning, of twenty-two feeding mule deer in various degrees of late-winter raggedness. A few had scars, possible reminders of the valley's heavy cougar population; one or two were limping and most were bony. They would arrive each morning just after dawn, the frost still white on the ground, in groups of three and four to feed. On our drive across the plateau a few days earlier, we had seen no deer, not one, and very few tracks.

I thought I would see how far I could walk, snow line permitting, up a good cow trail that our hosts, Iris and Dennis Redford, had described that led from above the old mill site near the top end of the lake to the north end of the Potato Mountains (Chunazch'ez). I'd heard and read that spring beauty (Indian potatoes) grew in the meadows at tree line up there. I wanted to confirm for myself that the stock trail was the same as the route taken by aboriginal people from Redstone and points north and northwest in the old days when they went up into the mountains on their annual trips to dig wild potatoes. As well I was curious about the location of a central camping place somewhere up there where all those old-time visitors to the mountains — not only from the northwest, but also from the northeast and east, the Chilko River and Nemaiah Valley — got together for a summer gathering.

So Tom is fishing for rainbows and I'm walking up this trail, a nice, steady, elevation-eating grade, the way stock trails usually are, with my eyes wide open and a dozen questions in my head.

I had been uncertain at first about the original starting point of the trail, as there seemed to be different opinions held by Tatlayoko residents as to where that point might be. In *Chiwid*, a collection of interviews with Chilcotin old-timers gathered by Sage Birchwater, Harry Haynes, who lived in the lower Tatlayoko Valley, remembered the Indian people going by his house to a spot where they parked their wagons and continued up the trail on foot and on horseback. Alex Matheson, on the other hand, in the same book, said their trail and base camp was on a wide bench where Ken Hesch, a local rancher, used to live, a couple of miles farther south. The only thing for me to do was to drive my Pathfinder as far as the narrow track would allow to an old cow camp and holding corrals partway up the lower slopes, get out, lace on my hiking boots and walk the cow trail up the hillside from there. I found

as I gained elevation that the open woods allowed good views of the valley bottom below and of the lay of the land up the slopes in front of me, and I became reasonably convinced that the aboriginal route and the trail I was walking on were one and the same.

Mainly, I just wanted to walk, activate the cardiovascular, get that mild, mindless, endorphin high that walking uphill brings and see what sorts of wild creatures had preceded me.

There was much to see.

It had snowed and melted at the lower elevations three or four days earlier. The ground was wet but fairly hard, with plenty of soft stretches where the many four-legged walkers could leave their marks. That's how trails are in early spring; with the snow line low, they become freeways. There were almost no deer tracks — the deer were all down at lake level — but a cow moose and last year's calf, browsers both, had come up. There were dog tracks, either a large coyote or a small wolf; I'm not an astute enough tracker to tell the difference. Dennis Redford said that wolves had taken horses right in the pastures that winter. And there were grizzly tracks, marked by their size, the length of their claws and the toes in straight rows (black bear toes curve around the main foot pad). There were four sets, two indistinct ones going up, and two

Iris and Dennis Redford's cabin at the north end of Tatlayoko Lake. The reliable Pathfinder, which takes me to all manner of wild places, is parked in front.

very clear sets coming back down.

The older, uphill sets had obviously been rained or snowed upon — the details and edges were rough, fuzzy, pock-marked — but the tracks, especially one set, were unmistakably large, unmistakably grizzly. These tracks did not wander but headed purposefully straight up the trail. The other pair of track sets, clearly made more recently by the same two bears, were more numerous and more distinct, the edges clear, the pad marks smooth, tiny creases and cracks showing. One was a good-sized bear. Her rear feet (I assumed it was a female) were over half again as long as my hand, her front feet two inches wider than my hand fully spread out. Each toe pad was as wide as my first two fingers together. The smaller tracks were slower to reveal themselves. This second bear had preceded what I assumed to be its mother down the hill; the mother's great foot pads tended to cover the smaller ones. I'd bet this bear was a cub from the previous year, travelling with its mother, and that over the winter they had denned together somewhere down the hillside, having only just come out of hibernation. Mama bears are known to cross country under cover of snow or rainstorms, and at night. The first grizzlies I ever saw, aside from people-familiar bears along roadsides in national parks, were

Some of the many deer that come to feed each morning in front of the cabin in the month of May.

a mother and cub crossing a road up off the Fraser by the Marble Mountains, west of Clinton, in an October snowfall in 1970.

Dennis had mentioned that there was a resident female grizzly, known to locals, who denned across on the unpeopled west side of the Tatlayoko Valley somewhere under Niut (Eniyud) Mountain, close to the lake. That's exactly where the downhill tracks were aimed.

I didn't expect the bears were still in this neighbourhood, but my attention stiffened anyhow, as if I was doing practice. I am sure that tens of thousands of years of paying attention in wild places — not too mindless, not too pointed, seeking sustenance or mere survival — is the origin of our efforts to refine our minds, otherwise known as spiritual practice, the world over. I found myself looking around more carefully for irregularities, turning to peer back over my shoulder occasionally, even sniffing the air. When we were out hunting when I was a teenager, scanning the skyline for watchful deer along old logging rights-of-way in second-growth up behind Port McNeill, my hunter father would say, "Look for what doesn't fit." And he would spot animals too, all kinds. Now I was getting a small shot of that old wilderness adrenalin, that edge that is wild. What makes wild is a thousand, thousand details beyond the laws of humans, from the smallest microbes in the forest duff and the dirt in our backyards to the largest Douglas fir, to bedrock itself. The subtle energies in this place had not retreated or gone dead yet; the presence of bear is a large part of why.

'Look for what doesn't fit'

I carried on and within a mile, just below snow line, I came to an exposed grassy point facing southwest, a sloping meadow about 200 yards long and 100 plus wide that, despite the elevation, had warmed up enough to encourage early plant growth. There the beginnings of a crop of yellow cinquefoils, hints of spring beauty, the new season's incipient balsamroot leaves, and tens of thousands of dandelion plants, each barely an inch high, not close to flowering yet, were just starting. All over, particularly among the dandelions, were the pad marks of mother and junior grizzly. All around and through the dandelions, and only the dandelions, the earth had been fastidiously, delicately, scratched and combed and the tiny plants eaten. Here and there, a few plant survivors were still standing, roots exposed, the fragile little stems a transparent white, looking like miniature stalagmites against the black earth.

I digested, with slow amazement, the contrast between the sheer size of mother bear's great paws and claws and the dandelion plantlets, and the fact that nowhere was the earth roughly disturbed, as bears, particularly grizzlies, are capable of doing. I have seen alpine meadows pock-marked with dirt piles and holes where they had moved large boulders to get at hoary marmots, for example. Here the two animals, in need of greens to loosen their winter's constipation, had gently scratched close to half the meadow. This must have taken hours, days.

I was feeling somewhat in awe

That bear mother had roused herself from hibernation, (calling her soul back from the dark season's circumpolar wandering, as envisioned by our distant, northern, hunter-gatherer ancestors) and come up the mountainside. Stories of Bear Mother are common to primal cultures around the entire sub-arctic world, from Norway to northern Canada, and go back to Neanderthal times if the repeated archaeological evidence of accumulated bear skulls and pollen offerings in European and Asian burial sites are valid indications. This bear mother travelled unobtrusively, with clear intention, to this precise place as she had done in seasons past, taught by her own mother no doubt. This was likely a practice as old as the remnant Coast Range ice fields. Ignoring the spring beauties' still-tiny corms, she got the laxative and a little of the nourishment she was after for herself and her young one. Then she had walked down the hill again, placing each of those great feet of hers methodically, deliberately, on the damp spring earth, her head swaying slowly, not seeming to hurry, but covering the ground. She would be nose up, smelling the air and watching for her offspring and for danger, her unassuming, moving mother eyes enduringly patient, unless threat was in the air. I would guess that she had timed her descent so that she could cross the Tatlayoko road in the safety of night.

I stood up from my study and musings and looked around me. The main trail carried on to the northeast, across the meadow, dark with the bears' scratchings, and up through the snowdrifts to a defile on the skyline that was the access to the ridge of the

Potato Mountains. At the far end of the meadow was another trail, not much used, branching away from the one I was on and leading straight north to a barely discernible entrance in a band of stunted timber. I walked over to check it out and found that the aperture was an old, hand-chopped connection to more sloped grazing meadows along the sidehill. I could see horses and cattle, not much bigger than specks, on open pasturelands below me, up-valley a bit. I scanned the slopes with my binocs and concluded that this track was not of major importance. I felt certain now that the old-time folks coming up from Redstone and points northwest on the Chilcotin plateau had used the same trail that I had just sweated my way up. Some may even have camped where the cow camp, quite visible down the hill below me, is now. There are tall willow thickets on that part of the hillside, so there must be a water source somewhere, a dugout seep maybe; otherwise how would corralled stock be watered? If the old-time folks were going up for Indian potatoes, they would go in late June or early July; there was sure to be runoff. Maybe there is a little creek farther south that I had not yet seen.

I was feeling somewhat in awe of the situation, the presence of bear mother pervading this place, and the lively story unfolding.

On top of Potato Mountain ridge after an early September snowfall, looking West over Tatlayoko Lake.

I stood quietly and mouthed a few words of surrender and thanks. "We are created and sustained, protected and surrounded by all this divine liveliness," I thought. "These are just a few of the many spokes in the great wheel," a voice in my innards said, felt, but not heard, "and it is one wheel turning."

A cold breeze was coming up off Tatlayoko Lake. The great mountains to the west and south, down past the bottom of the lake, in the heart of the Coast Range, were thickly blanketed in snow. I turned and walked back down the hill. There were no obvious signs of trails merging anywhere.

'It's the bear's bedroom'

Just uphill from the corral, I remembered that the first time I'd been up this trail, this particular stretch of hillside in fact, I was with Marne in late July 1996 and we had not walked far. All around us among the birch and willow clumps, the Douglas fir stumps, the soopolallie bushes, the large-leafed burdock and cow parsnip, were old bear poop plaps, gruel-thin, soopolallie-berry red, dozens of them, hardened by the sun's heat and scattered over a large area. Some bear over many days, driven by a need to purge itself, a need for the taste of bitter, had been eating soap berries and more soap berries, and they had passed right through, "like shit through a goose," as the saying goes.

In the weeds and tall grass there were a number of flattened-down spots as if something, bear-sized, had been lying down here and there over a period of time. The bright green of new grass showed in some of them. "It's the bear's bedroom," Marne said, a little nervously. On the bare twigs and branches of some willows and birches around were hundreds of bright white butterflies, unmoving, each the approximate size of a silver dollar. A few lay on the ground, obviously lifeless. Marne was edging her way down the hillside.

A month earlier, two miles farther south, Sven Satre, a local rancher out looking for cattle, had been spooked off his horse. Searchers found him the following morning, partly eaten and guarded by a large black bear. Big male black bears will, very rarely, hunt humans for food. This one had already been recognized locally for his conspicuous lack of fear. When sighted by people, instead of running, he would occasionally advance with no hint of tentativeness toward them. To curl up and play dead in front of such a bear, in stalking mode, would be a serious mistake, the equivalent of presenting oneself as fresh meat, pre-wrapped, in a butcher shop. The bear's tracks on top of those of Sven's horse indicated a running chase along an old treed-in logging road. The horse was found some distance away, with saddle cinch freshly broken, anchored by the halter shank still connected to the saddle. The bear was promptly shot. Given the wide range of a typical male black bear's territory, it is possible this was the same bear as the one at the soopolallie slope.

There was, of course, much shock and mourning in the community. And some rage. One man, a well-known local "home guard," went on a minor rampage, killing a couple of black bears unfortunate enough to wander through his part of the valley for retribution and the ritual of it.

I caught up to Tom on the main Tatlayoko road. He was walking along, feeling kind of proud of himself, with his fishing gear and a twenty-two-inch, four-pound rainbow trout. He had caught six, kept the first and carefully released the rest. Knowing there was lots of bear activity around, Tom told me that he carried the big trout by hand rather than in his pack. That way, he could drop it quickly if necessary. We ate that fish, cooked expertly by Tom, for supper, joined by our hosts, Dennis and Iris. I'd set aside a decent bottle of wine for the imbibers in the group in anticipation of a pleasant evening.

Four months later, in September, I was back, travelling alone. I wanted to see the Chilcotin in the fall, something I hadn't been able to do in decades. I rented Iris and Dennis's trailer at Tatlayoko Lake for a week at a reasonable rate and did some day walking.

Iris had told me about another cow-trail access into the southern end of the Potatoes. It started above a corral by the road that ran along the lake into Bracewell's place. I'd been wanting to explore that area for a while because, beyond a love of the alpine land, especially in autumn, and general curiosity, I was fairly sure it would take me near the Chilcotin (Tsilhqot'in) people's old-time gathering place.

One afternoon I checked out the first half-mile to make sure I could find the route; there were two or three trails crisscrossing the area. The following morning I took off. I was bound somewhat by time; I had gotten a late start, and Iris and Dennis had invited me for dinner that evening. Deadlines can squeeze you sometimes.

I was slightly wary of bears; there were lots in this neighbourhood and I did not want surprises. In June, when the snow is low, the area is plugged with them. Peter Stein and I counted eight black bears one time, about one a mile, on a single road, the same road I would be travelling, a stretch I've taken to calling bear alley. This route is along a warm slope above Tatlayoko Lake with plenty of food sources, brushy thickets, grassy road margins, berry-bush hillsides and immediate access to truly wild country down the Homathko River and into the main Coast Mountains.

The trail began at 2900 feet, went steeply uphill, veered to the east and levelled out through dark standing timber for a mile and a half. After passing a rough pole fence and corral above Bracewell's spread, it turned sharply uphill again and almost immediately took me up a grassy nose onto beautiful south-facing open slopes with bright ribbons of poplars and willows in autumn oranges and yellows up and down watercourses, now mostly dry. I noted patches of dead balsamroot sunflowers, not a surprise on this warm hillside, but growing much closer into the cold, dark, wet Coast Range than I had previously seen. These slopes would have been a sea of golden faces in late spring.

The trail was dry, a fine, fine dirt; naturally I looked for animal sign. I saw tracks: a few deer, a cow moose with calf, cattle, cowshit, one or two piles of old, nondescript bear poop, and joining the trail about halfway up, from downhill to the west, a black bear mother with a very small cub, an outcome of the previous late spring perhaps. That was one cub, for sure, who'd be sleeping in his mother's den in the cold of the coming winter.

It was a steep trail now, switchbacking forth and back, across and up the hillside, and I got a good sweaty workout. My face was right into the trail and I could see the bear tracks clearly; they were very fresh, well-defined. I could see every crease; the fine dirt held the details well.

By high noon the topography was semi-levelling out into rolling grassy tundra with dry rills and rocky tarns and dark green clumps of alpine fir. The frost-touched grasses and foreshortened willows were various shades of red and golden.

As the gradient decreased, the bear tracks promptly veered off along the top of the drop-off and tree line to the east. No doubt mama was heading for some bear delicacy or other. According to those who know — the Craighead brothers, Doug Peacock, Stephen Herrero, Andy Russell, etc. — bears, black or grizzly, can vary greatly in their personal food preferences, just like people. It looked like this bear mama knew exactly where she was going and had used that cow trail for help in getting up the steep parts.

Despite my initial caution, I saw no bears on this walk. I had not taken into account the dry summer and the failed berry crops in the high country. Also, some of the local black bear and grizzly populations were likely over at the Chilko River cruising for salmon bits left over from the big sockeye run there.

> *A person could get quite light-headed up here*

I stopped to eat lunch and drink in the mountain air, like a particularly good wine you prefer to savour. That fine fall atmosphere was palpable, and a person could get quite light-headed up here. I could see down to the bottom end of Tatlayoko Lake, less than 4000 feet of elevation below me, and the Homathko River dropping down through the big Coast Mountains, past the biggest, Mount Queen Bess, and all the rest. The great Mount Waddington was not far away to the west-southwest, but, as usual, it was out of sight.

There were old mine claims and sites down there on the slopes of Mount Moore, off Nostetuko Creek, worked on and off over the twentieth century. Iris Redford (formerly Moore) had been employed there as a cook. "It was isolated," she said in her usual matter-of-fact manner.

Harry Haynes had claimed there were bighorn sheep, Californias, down the Homathko and the lower end of West Branch, a remnant maybe, like that tiny band still surviving a long way north, in a windblown corner of the Ilgachuz Mountains by Anahim Lake. I have to say, that country downriver has always seemed to me to be too coasty, too thick-timbered and wet for sheep country. Iris says wet-coast cedar and hemlock grow along bends in the river bottom not far down. But Lee Butler over at West Branch, a man born to the woods, told me recently that where the Homathko River veers west there are dry, open, rock slopes facing south, and a small herd tended to locate around there in earlier days. I would be curious to get down there to see that area for myself, especially as Murderer's Bar, the site of the major conflict and loss of life in the Chilcotin War, is down that way. Alfred Waddington, a Victoria businessman, attempted to build a road to the central BC interior from Bute Inlet in the early 1860s, but relations between the newcomers and local Tsilhqot'in people were disrespectful and difficult, culminating in violence. Fourteen members of the work crew were killed. Three or four further conflicts occurred over the following summer, and five more people were killed, including the infamous ex-Hudson's Bay man Donald McLean, who had once been factor at short-lived Fort Chilcotin, situated near the confluence of the Chilko and Chilcotin rivers. The Tsilhqot'in peoples of the Chilcotin country demonstrated little interest in the fur business.

I walked up onto the ridge for a short mile and looked around. The country was resting, re-creating in the late September calm. On the way up I'd encountered a couple of deer; several raspy Clark's nutcrackers who, true to their deepest corvid natures, informed the world of my intrusion (you get away with nothing up here in these high places); soaring eagles attaining a large view of things; noisy ravens, coyotes of the sky, playing cannily in the little updraughts off the lake; a quick sharp-shinned hawk on pressing business, cutting the air, leaving no wake; blue grouse stalking away, maintaining their dignity; blurry bumblebees; and a background of grasshoppers, clicking and snapping, to add to the feel of things.

The ridge gradually gained elevation as I walked, rising to about 6900 feet. From the edge it was a near-straight drop to Tatlayoko Lake. Except for the ridge rising to the north, I could see all around me. I got the small- and large-scale maps out and studied them, taking the time to match the contour lines, relative positions, distances, elevations and drainages on the maps to what I was seeing around me. Most of the big mountains to the west and southwest and south were visible, a snowy, rocky wall of them: Niut, the backside of Razorback, Blackhorn maybe, Ottarasko, Success, Reliance, Homathko Peak, the Anniversary Range, including, of course, Queen Bess, Mount Moore, and the Liberated Group. Whitesaddle was out of sight. There was much snow, much ice, and sunlight glinting.

To the east was the eastern ridge of the Potatoes; farther east

were peaks somewhat lower than those in the west: Tullin Mountain, Mount Nemaia, Konni (Xeni) Mountain and, overlooking Nemaiah Valley, massive Mount Tatlow, now known by its original name, Ts'yl-os, 10,058 feet tall, *axis mundi* to the Xeni Gwet'in, the people who live there, and a living, truly commanding presence. Don Brooks and I walked up Ts'yl-os in stages in 1986. That was a memorable trip, notable for lots of up, food poisoning, horned owls calling to each other across an open creek bottom at dusk, a solitary blink from a single light far out on the dark plateau, and sunlit wild horses of all colours, tails twitching on a cool bright morning high on the alpine slopes of Ts'yl-os. It wasn't that far over to Chilko Lake — five or six miles maybe — and there were bighorn sheep for sure, there in the dry Nemaia Mountains on the east side of the lake.

On top of Potato Mountain ridge after an early September snowfall, looking east over Chilko Lake at the Nemaia Mountains.

It is considered poor manners, and dangerous, apparently, to stare or point at Ts'yl-os, but we did not know that then. Possibly it was one of us in a moment of thoughtlessness who caused the insult and the touchy stomachs. Don dry-heaved all night. I was weaned and raised on upcoast swamp water, at times brown and fishy in the fall, and did not fare so badly, though I too felt exceedingly feeble and unwilling to eat.

I lay on my side in the heather, taking in the mountain pan-

orama on the far side of Tatlayoko, and after a while, before suc-
cumbing absolutely to the wine-air and oozing permanently into
the thin alpine earth, I stirred myself, folded up my maps, rose
and, very slowly at first, walked on north. I got to a point just short
of the so-called Tilting Lakes. They were barely visible, looking rather shallow on a wide, treeless flat in front of and a little
below me, above the headwaters of Fossil Creek. Looking east,
I could see that it was a good hour's walk from there down into
the Echo Lakes, where the old-time Indian summer camps would
most likely be. I'd been told that if I walked along the ridge of the
Potatoes another three miles or so, to the highest point, 7200 feet
up, I could look right up the Jamison Creek drainage, like looking up a big gun sight, across Tatlayoko Lake to the west. From
that point, word has it, I would be able to see Mount Waddington, 13,177 feet or 13,250 depending on who's counting, the highest
mountain in the Coast Range, the highest mountain entirely in
British Columbia. Mount Waddington is a mountain very few of
us ever see, because it is so closely surrounded by all those other
great mountains in the 10,000- to 12,000-foot range, Mounts Tie-
demann, Munday and Asperity, Combatant Mountain, the Serra
Peaks. Seeing Mount Waddington is awe-inspiring, something
like an audience with the Dalai Lama or the Pope for some, or
seeing a glory.

A glory is a rare meteorological phenomenon produced by
reflected light from the edges of water droplets. I experienced a
glory once after a sharp, bright rain on the top of Mount Arrowsmith on Vancouver Island in my young and agile years. It was a
marvel to see our own shadows outlined on Cokely Ridge across
from us, made huge by the magnifying effect of low, late-afternoon sunlight through billions of tiny wet prisms. The sun's rays
radiated golden behind us and around our shadows on Cokely like
auras. Rainbows were intense, the background sky near-black.

Phyllis and Don Munday saw Waddington in 1925 from Arrowsmith, a very long way south of that great peak. They called the
unnamed eminence Mystery Mountain. (The clean air in those
early years allowed them that first enticing view. We could not see
Waddington from Arrowsmith now.) The sighting was a precursor
to their many approaches to climb the Waddington group. At that
time, the Mundays and their colleagues had plenty of persistence

and stamina, not to mention audacity. Most of their expeditions took them all the way from tidewater at the head of Bute Inlet to the snow and ice fields in the core of the Coast Mountains, carrying sixty-pound, or heavier, packs through some of the worst and steepest salal and mountain azalea jungles on the coast.

Don Brooks and I saw Mount Waddington once, in 1988. We flew from Nimpo Lake into "Brooks" Lake, which is close to Wilderness Mountain, farther north in the Coast Range, at the headwaters of McClinchy Creek. Don is the only guy I know personally who has had a lake named after him. This honour, probably not official, was conferred upon him by the pilot, for whom landing on this particular lake was a first. We did a low flyby to look for rocks, then circled and landed. The water's surface was flat-calm. "Brooks" Lake is a small lake without much takeoff room, and I expect the pilot has never gone back. We stayed out there and explored the country around in the cold — and I mean cold — with the dense, dark, granitic Coast Mountains closing in on us, for three days (there are no wide-valleyed Rocky Mountain vistas beckoning you onward out here). Then we walked out. We had to walk. Seaplanes require a longer run for take-off than for landing, and there is no way a plane, not even a Beaver, could have lifted off that short lake with our three heavy bodies and two sets of packs aboard.

Seeing Mount Waddington is awe-inspiring

On a ridge up Wilderness Mountain, where we were joined by a couple of uncharacteristically nervous mountain goats one sunny day (we were slightly above them), we got out the compasses and maps and lined up Waddington. From that angle we could actually see it. It sat there with authority palpable even from forty miles away, more lofty by far than the lesser mountains around it, so much so that it had its own weather system, a ring of clouds glued to the peak, as if it was a great cloud-magnet.

Our unsettled goat neighbours, after getting up and lying down several times, finally stood up and left, carefully picking their route down through the rocks in their usual deliberate manner and disappearing into some mysterious goat hole in the mountainside, the way goats seem to do. Despite the horseflies and the chill, we, on the other hand, considered ourselves blessed.

But up there on the Potato Range, I was running out of time. I knew I would be up here again to check out routes and details as soon as possible, maybe the next year. I turned, reluctantly, and started off down the ridge, stopping every quarter mile or so to take it all in, to get it, to inhale it, to stay in it, even to try, futilely, to keep it, as if I could stuff it into my backpack. The good thing about walking downhill, apart from the fact that it is easier on the heart and lungs, is that when you are not selecting places to put your feet, you are looking out at the mountain view, especially in the late afternoon when the sun is lowering, the glaciers blazing, the valleys darkening, the shadows creeping, the night's cool edge is moving in, the small creatures of dusk are starting to rustle, and the willows and grasses and stubby little poplars are that much more deeply golden than in the bright light of midday.

By the time I rejoined the black bears' tracks at the top of the steep, the shadows were long. I stopped and sank to my knees to examine the footprints closely. I put my face right up to them. A little breeze, a zephyr, stirred through the dead arrowhead-leafed sunflowers, making a dry rattling sound, quite strong, insistent. The cub was very young for so late in the year. His tracks were so small, barely four inches long and triangular. The heels of his rear feet came to little points, like the classic, stylized bear tracks in some pictograph panels on select rock faces down south, on the great panel at Seton Creek, for example, or up the Stein River.

I looked at the little bear tracks more closely than I've ever looked at animal tracks. In the slanting, late-afternoon light, with the shadows in them like tiny veins giving definition and life, I could see the little creases, minute, crossing and crisscrossing each paw, myriads of them in the fine dust. I could see them as clearly as I see the creases in my own paw-hand.

FIVE

Spring Beauty

— say the names say the names
and listen to yourself
an echo in the mountains . . .

say the names
as if they were your soul
lost among the mountains
a soul you mislaid
and found again rejoicing . . .

'SAY THE NAMES'

AL PURDY

Like many, I seem drawn to wild: wild places, wild in the rocks, the swamps, wild down off the high slopes, wild in settled places and around our efforts to survive, wild people even. We arrive, we clear land, we build and grow, we age, die and our worlds crumble, our fields grow in and wild comes again creeping up between the cracks. I've seen wild places all my life, but what I'm really after is the wild that is in all things, the wild that is the core of my own being. I am not alone in this. Some wild people impel us to look and see: "Wild" is another word for *Tao* or *Dharma*. The way. Empty. Real.

In 1995, New Star Books published *Chiwid*, a truly wonderful collection of interviews with Chilcotin old-timers gathered over many years by Sage Birchwater. The book is full of detailed information and insights about Chilcotin life in the old days. Above all, it is a look from many angles at a Chilcotin legend, Lillie Skinner, a wandering woman, born about 1900, who, after a radically violent and traumatic incident in her young adulthood, took to the woods. She followed the old seasonal survival rounds, fishing, hunting and gathering across the plateau from Anahim Lake to Toosey and back. She became known as Chiwid, "Chickadee." Chickadees are the little birds you hear often in winter, in the Chilcotin and other northern parts, perkily going about their business. Chickadees survive in small groups, almost miraculously, through the coldest weather, unlike nearly all other small birds, which move south. Like the chickadees, Chiwid lived outside all

The author with Sage Birchwater, Williams Lake, 2006

winter, sleeping sometimes in snowdrifts, feet hard as ice, seeming to prefer that to a heated cabin where, no matter how much wood she put in the stove, she could never get warm. Illa Graham, in *Chiwid*, said, "She was like a little bird, like the ptarmigans that would sleep right in the snow. That's what she used to do, too. No one knew how she did it."

Over the years, as people saw her walking, summer or winter, through one place or another, or came across her little camps in the middle of God knows where, her mystique grew. Chilcotin people, aboriginal and settler, began to look for her and, as she grew old, keep an eye on her. In her last years she camped close to relatives at Stone Reserve. I like to think I saw her camping place in the early 80s, not long before she died — a ragged collection of blankets draped over a fence, and perhaps a pot, on the outskirts of Stone. Or maybe I'm just kidding myself. Chiwid is a story in herself, and it would take a full reading of the wide array of perceptions in Sage's book as to who she was and why she did what she did to even begin to understand her.

Like the Hoofprints in History series out of Tatla Lake School, Veera Bonner's book *Chilcotin: Preserving Pioneer Memories*, and aspects of Paul St. Pierre's work, much of which was fiction, but often based closely on real individuals, places and situations, *Chiwid* has huge authority. The older folks interviewed are talking in their own words, in detail, unfiltered by outsider interpretation about what they did when and, especially, where in the old days. So much of movement then was on foot or horseback, or by wagon, and the passage of time would be slow, the landscapes varying. There would be opportunities and cues for precise observation, memory and contemplation of season and place.

A reference in a Paul St. Pierre article in *Chilcotin Holiday* to Chiwid's half-brother Ollie Nikolaye, and conversations with Veera Bonner, led me to read *Chiwid*. It was Veera who took the fine photograph of her that graces the book's front cover. Chiwid had been camped in the willows on the edge of Fletcher Lake near Big Creek, visiting her daughter Juliana Setah, who was living across the lake at the time. After I had read the book carefully, I wrote to Sage Birchwater expressing my deep appreciation. I addressed my letter to Sage, c/o Anahim Lake, trusting it would reach him eventually. It was mailed on July 2, 1996. After a wander around the West Chilcotin for the rest of the year, the letter reached Sage on Christmas day and he responded by mail soon after. He and I have become friends.

Incidentally, I sent a copy of *Chiwid* to Gary Snyder, the American poet (or should I say "Turtle Island poet") and long-time student and teacher of Zen Buddhism, and suggested it as a Christmas present for Peter Nabokov, son of Vladimir, a prominent anthropologist. Both were extravagant in their admiration of what *Chiwid* accomplished. "It is truly special, crosses all kinds of spiritual, shamanic, ethnic, cultural lines," wrote Gary. Nabokov read his copy on Christmas Eve and was so excited by it that he requested five more copies from his daughter, Fiona Neufeld in Victoria, the next day.

In Chiwid's wanderings, she came often to the Tatlayoko Valley and Big Eagle (Choelquoit) Lake area. One daughter, Mary Jane, married into the Lulua family and lived near the west end of Big Eagle. Another, Juliana Setah, and her husband, Willie George, lived and worked for a while at Skinner Meadow, feeding cattle in wintertime for K.B. Moore before moving east to Big Creek. Chiwid often passed through. Skinner Meadow is named after Chiwid's father, Charley Skinner, who came into this country from the United States with big dreams and good horses, large numbers of which eventually went wild. Little is known about her mother, Lauzap, except that she was deaf and unable to speak and that she gave birth to several other children. Sometimes Chiwid's ranging would take her into the Potato Mountains (Chunazch'ez) to dig spring beauty and maybe hunt deer. The old-timers say that before her sight went, she could bring down a moose using her single-shot .22 calibre rifle, with a bead of balled-up spruce gum for a front sight. Some say her accuracy was good even after she lost her visual acuity.

Spring beauty (*Claytonia lanceolata* Pall ex Pursh.) is known locally as soont'ih, or wild potato, or Indian potato. It is a perennial herb about six inches tall, grown from an underground corm that is typically the size of a small cherry. It grows in masses in moist meadows at timberline and in higher alpine grasslands. Spring beauty flowers are white with pink veins, have five petals and grow in clusters of a dozen or so. From mid-June to mid-July, certain hillsides are white with Indian potato blooms. The relatively warm, outlying Potato Mountain range is the northernmost area where they grow in large numbers.

Spring beauty corms are starchy, edible, nutritious and accessible for digging. Traditionally, since "time immemorial," as the saying goes, they were dug with a small digging stick and laboriously collected by the basketful. The corms were steamed in underground pits and stored in similar stone-lined pits. They were a great source of carbohydrates for the old-time folks and were often used in trade with people from farther north.

She could bring down a moose using her single-shot .22 calibre rifle

Chiwid was not alone in going up into the high country to dig wild spuds. Until the 60s and 70s, Tsilhqot'in people in numbers from Redstone and points west, and from the northeast and east, the Chilko River and Nemaiah Valley, climbed up the ancient trails for potatoes every early summer. The Lulua sisters, Doris and Madeline, from Tullin Ranch below Tullin Mountain near Chilko Lake were still going up there as recently as fifteen or twenty years ago.

As mentioned in "Creases in a Bear Cub's Paw," there is a chapter in *Chiwid* on the Potato Mountains and the older residents' recollections of the aboriginal people passing by, on horseback and by wagon, down through Tatlayoko to the start of the mountain route. Here they would park their wagons, load up their packhorses and, men, women and children, carry on. This passing was momentous, both for the participants and for the watchers, and it is easy to feel the sense of occasion, and even awe, in the words of interviewees. Alex Matheson said, "I never went up there myself, but they used to come by my place in a big gypsy convoy of wagons and horses. Whole families would go up there. All the old Indians and kids. Kids would be riding all alone on a saddle horse following a wagon. Often they weren't much higher than the saddle horn itself." Bud McLean, from Chilko Lake, referred to those years as "Really the glamorous old days … A bloody caravan went in there." Doris Lulua from Tullin Ranch says, "Every year my mother went up Potato Mountain … People just live on deer and potatoes. Them days no welfare. Had stampede grounds. Mountain race. Lots of camp." Until they died recently, Donald Ekks and his wife Emily lived at Cochin Lake. Donald recounts,

"Make'em horse race. All Redstone, Nemiah, ... Anahim Lake.
Somebody bring homebrew ... Sure lots of fun. My woman dig 50
pounds in one day. Long time ago."

Clearly the people went up the mountain for more than wild
potatoes. They went up there to rodeo. They hunted deer and
marmots, feasted, socialized, caught up on the news, made homebrew, played the stick game lahal, wrestled, made bets and raced
horses. Somewhere up on a high flat above tree line, supposedly
in full view of the great snowy mountains to the west and south, so
the tales go, there was the fabled racetrack. That would have to be
one of the most fantastic racetracks anywhere.

For myself, I first saw that country in 1968–69, about the time
the "gypsy convoys" were ending. Like many, I recall vividly the
people travelling about the country in their old rubber-tired wagons, or camped across from Stuart's store, now long closed, at Redstone and down the road from Lee's Corner, their white canvas
tents flapping, horses tethered in behind, the blue smoke rising
from their kitchen fires. I remember, in the early fall of 1970, going
into Red Brush, north of Puntzi Lake, along a pothole-pocked
dirt road, with John Rathjen — who subsequently drowned in
Alkali Lake attempting to save the life of a friend — and seeing
the people camped out in the backcountry, cutting meadow hay,
shoeing their horses, gathering their cattle, minding their babies.
I remember Old Michel standing in his open woodshed somewhere in that Redbrush country, the cold October wind blowing
through the dry, yellow grass, his horses grazing down near the far
meadow edge. He was slowly, very slowly, cutting firewood with
a blade and wooden maul, and even more slowly saying, "Yeah,
[pause] I been here aaaaall my life," stretching out the adjective
in that old-time, drawn-out, singing way to indicate just how long.
I recall old memory voices like that from old-generation folks in
my young childhood up the North Thompson River after the war.
Old Michel was reputed to be a hundred years old, give or take, at
that time. That would mean he was born within a very few years
of the Chilcotin War. No doubt some of the people involved were
his own relatives, and he would have known them. He died within
the year: "he did not winter," as the people say.

Increasingly I was imbued with a desire to explore the whole of
the Potato Mountains massif, the eastern as well as the western

Barry Menhinick above Lorna Lake, upper Big Creek, on the "epic" ten-day, South Chilcotin horse trip of 1997.

ridge, and to confirm for myself the location of the old trail from the northeast. I had been curious for some time to find the main camping place, the "stampede ground," where the people met. Bud McLean said it was "at a flat between two lakes," but there were three pairs of lakes in the Potatoes, and it was a fair-sized piece of territory to cover, especially on foot.

My long-time friend and co-hiker Don Brooks and I had done some kind of extended backpacking or riding trip together each

The mighty Smokey and Jake at Warner Pass, 7800 feet up, between the Gun Creek and Taseko River drainages.

year but one for twenty-three years at that point, an accomplishment in its own right. We had gotten to where we almost didn't need to talk. On a near-epic (for us) ten-day, hundred-mile horse trip, not including layovers and day trips, in 1997, with Don's wife Dana and my partner Marne, outfitted by Barry Menhinick out of Goldbridge, we rode in a great circle from the top end of Relay Creek, through the South Chilcotin Mountains and back to Relay again. We camped in the headwaters of Big Creek below Lorna Lake and did a day trip to the head of Tosh Creek and up onto the wild and wide open Dil-Dil Plateau. Then we rode up Grant Creek and over Iron Pass to Horse Heaven below the Battlements in the upper Taseko River drainage, over Warner Pass (7800 feet up, more than 500 feet higher than the highest peak on Vancouver Island), down and up Gun and Tyax Creeks, and over Deer and Elbow Passes to Graveyard Valley before returning to Relay Creek. We got to know Russ Dreger, Barry's wrangler for the trip. Russ is from Knutsford, near Kamloops, and has a few horses and ideas of his own. I rode at the back of the train with Russ, and over a trip of that duration we had lots of time to talk. The particular idea that we mulled was a "walk and horse pack" concept. Over the winter I ran the idea by him of trying it out for real in the Potato

Mountains, and he agreed. We would get together in late August. Don and I would walk; Russ and his horse pal Gerry Nagle, also from Knutsford, who had his own horses, a big horse trailer, and a cheerful outlook, would ride and lead packhorses with all our cooking and sleeping gear, or as Russ puts it, "With all the necessities of a very fine camp as well as some of the luxuries that aging backpackers [that's Don and me] tend to leave behind."

The route we sussed out over that winter on a couple of 1:50,000 topographic maps was down the eastern ridge from Chilko Lake via Tullin Mountain to a pair of lakes — the most likely pair for the Indian gathering place in my view — at the south end of the massif where the two ridges of the Potatoes come together above Cheshi Pass. From there, we would turn north along the main ridge, in full sight of the snowcapped mountains in the heart of the Coast Range and Tatlayoko Lake, and come down, eventually, into Tatlayoko on the old trail at the north end of the lake above Iris and Dennis Redford's place. We would be keeping a lookout all the while for signs of the traditional aboriginal meeting place and, hopefully, the legendary racetrack.

It was an interesting trip, requiring considerable time and preparation to manage parking the horse trailer, arranging to leave a car for our return with Iris, organizing the gear, packing the horses, etc. Our preparations were punctuated by a visit from Elizabeth Schuk, a good-looking cowgirl who rode a white horse as if she was glued to it, a cow dog in training named Dooley (at first, I thought his name was sonovabitch), and about 80 or 90 reluctant, irritated, bawling cows, "black-baldies," Angus-Hereford crosses, that she was moving past our base at the old-time campsite at Big Eagle Lake. Dooley got a kick out of running those cows in any direction but the right one, which was why all that extreme cussing was happening. Part of the problem, also, was that Russ had put his horses in the corral next to the gate that Elizabeth was intending to push the cattle through. We quickly moved the horses out of there.

Elizabeth was staying down the lake at the old C1 Ranch line cabin, and later, after she had joined us for dinner, beer and horse talk — "Everybody loves a buckskin" — she was visited by a couple of boyfriends in their pickup truck. Not long before dusk, a lone tourist in a camper with California plates showed up and took it upon himself to park under some poplars in front of the gate linking two grazing areas up the slope from us. The spot was pretty and had a fine view, but it blocked the only route for Elizabeth's friends to return. We retired for the evening, a little curious as to what would unfold. Return they did, just as we were falling asleep. We were shocked awake by the sound of their pickup high-revving, followed by a prolonged silence that climaxed in four loud gunshots, carefully and slowly spaced, then the sound of the pickup noisily pushing and banging its way up the sidehill, wheels spinning, past the camper and back onto the road and away.

There was nary a peep from Mr. California, just his window curtain twitching. We awoke before dawn to the sound of him quietly driving away, the sound of those shots no doubt still ringing in his ears, freshly educated, for better or worse, as to back-road etiquette in the Chilcotin.

The actual walking began propitiously on a pleasant morning with Don and me striding manfully up the northern flank of Tullin Mountain, overlooking Chilko Lake, minus our backpacks and flagging the route occasionally, as necessary. Little did we know that, behind us, one of the packhorses, Dillon by name, a project of Russ's, would get broncy and decide to put on a rodeo of his own. The result, as we were to learn at the end of the day, when we

Packhorses at Warner Pass in the South Chilcotin Mountains. My horse, Lady, a Dutch warmblood, a goer and fun to ride, is on the far right, planning new ways to be in charge.

met up and made camp, was the loss of an essential gas-stove fitting. This was important because, although we had no aversion to campfire cooking, the weather had been very hot and we were in fire season. Campfires were best avoided, if possible. Don cooked a fine meal that night over a carefully watched open fire. And the consequence of Dillon's rodeo was that the next morning Gerry had to ride his big gelding, Comet, all the way back down the mountain to look for, and eventually find, the missing part. We didn't mind too much; Russ stayed with the horses and spotted some clean grizzly tracks, a mother and young one, in the soft mud at the edge of the little pond where we got our water, and Don and I went walking. We went for two walks: one to check out our route beyond us to the south and to take pictures of Don with a bottle of Blue Mountain pinot noir, I believe, for the winery, the blue mountains behind Chilko Lake as a backdrop; and the other, a short walk into what I presume was Bud McLean's camp at Fish Lake. This is only one of at least two dozen "Fish Lakes" in this province; the old-time namers of places, too busy on their way to somewhere else, and surviving, frequently lacked imagination.

The ceiling, the next morning, was covered in high cirrus, and there were rain clouds in the far west. The light was bright and the day proceeded without incident. Our walk along the ridge was incredible, with views of Chilko Lake right up to its far end, forty miles away, and Tsuniah Lake, Konni (Xeni) Mountain, the Nemaiah Valley, Ts'yl-os Mountain at the centre of it all, and the great snow mountains of the Coast Range to the south and west. Without heavy packs, Don and I fairly floated. At the southern brow of the massif we were crossing, we could look west to the main ridge of the Potatoes and the pair of alpine lakes just below, called "Tilting" by some locals. These lakes are unnamed on all the topo maps, although a fairly recent Forest Renewal map refers to them as Dunlap and Gillian lakes. Whatever their name, they were no place to camp, now or in traditional times, being exposed and woodless. The main ridge looked very rideable. We picked our way carefully down through the krummholz brush and sidehill swamp into the partly timbered valley below, with the two main "Echo" Lakes, and Bracewell's mountain cabin between them.

We were walking along the margin of the first lake rather fast because Russ, Gerry, the rowdy Dillon and the rest of the packhorses had caught up to us again and we were keeping up with them. I was noticing the odd old, deep firepit here and there, and it was dawning on me that this was likely the main camping area I'd been curious about for so long. There was still a third pair of lakes at the north end of the ridge to check out, but this had to be the place.

Unfortunately, with the business of unpacking and tethering the horses, making camp, preparing dinner, drinking a can or two of beer, battening down the hatches for the rainstorm moving in, plus poor planning, I didn't give myself much time to prowl around. I did take a quick look up on a treed knoll behind the cabin and saw signs: more of those firepits and deep rock-lined storage pits, and what seemed to be the alcove-like windbreaks that might have shielded the campsites, which Petrus Rykes, one of the managers of Eagles Nest, a resort at Anahim Lake, had talked about. I looked north across a grassy flat to the second lake. There were half a dozen cattle on the sidehill. Slanting up the slope in the direction of the "Tilting" Lakes on the alpine bench above was an old trail, worn in and so steadily rising it looked as if it had been surveyed. I really wanted to do a more complete explore of the area.

Russ and I did do a short reconnoiter of the south end of the lake we were camped by, to be sure of the trail down into Bracewells should we need it. It was hard to miss. We encountered more cattle, a nearly infinite number of cow pies and considerable erosion at the outlet of the creek. The place had that barren, burnt look of overgrazing; there were tall, dried-out, fuzzy-leafed great mullein everywhere. Even Russ, who had put in his time cowboying and was a lifelong student of horse history and culture and a true aficionado of cowboy ways, said fervently in a big voice to no one in particular, "I hate cows in alpine meadows." Amen.

We awoke the following morning to find ourselves thoroughly socked in, rain, zero visibility, the whole shebang. My fantasy of a leisurely saunter along the ridge to the north end, gazing serenely around at the mighty mountains to the west, had pretty much evaporated, especially as we had a deadline and lacked time to wait for the storm to pass through. We decided to take the sure route, the wide stock trail downhill to Cheshi Pass, and then catch the road north along Tatlayoko Lake to Dennis and Iris's place.

Don and I took off; the horses and riders would follow. We soon passed by Bracewell's grand lodge overlooking some fields in the pass, with Alf Bracewell banging away at something solid and metallic in his machine shop. Alf was the catskinner who cleared the uphill end of the legendary, locally built road down the storied "big hill" to Bella Coola back in 1952 and is something of a legend himself.

We hoofed out the rest of the fifteen miles in record time; actually getting there ahead of the horses. Don has exceptionally long legs and is very rugged for a scientist. On our arrival, Iris poured us tall glasses of ice water fresh off the glaciers, beads of condensation on the glass and actually tastier than beer, it was that good! We waited on her now-sunny front stoop for the horse train to catch up. How many times have we walked out of a high-country rainstorm into low-elevation sunshine at day's end? As the old country saying goes, "If you don't like the weather, wait an hour or two."

'I hate cows in alpine meadows.' Amen.

But I was left with unfinished dreams. There was still that last pair of lakes to check, and I had not yet seen an Indian potato.

So two years later, Peter Stein, Sage Birchwater and I got together to walk up that northwest trail from Tatlayoko into the Potatoes to find spring beauty and finally see them in peak bloom and full corm. It had been a heavy winter that year; the snow was slow leaving. The mountainside was soft with meltwater. We started early, just ahead of over a hundred cows and calves that had been collected at the old corrals near the bottom the night before, disrupted and perturbed as usual, to be driven up the steep hillside to their summer range.

We enjoyed the walk up the mountain in the bright morning air, yakking and gawking, working up a sweat and covering the nearly 3000 feet of elevation comfortably. We came out soon, at the upper edge of the alpine fir and stunted lodgepoles, into wet meadows, white with spring beauty, green with spring. Along the ridges, whitebark pine rose tall and bushy-topped above their neighbours, dominating the skyline.

Whitebarks are distinguished from the smaller lodgepole pine by their long needles in clumps of five and are usually striking trees to see: they live near tree line (over 4000 feet at this lati-

Russ Dreger, Don Brooks and Gerry Nagle on the eastern ridge of the Potato Mountains in 1998. The broncy Dillon is the horse on the right with his ears back.

tude); their spreading branches and multi-trunks provide cover in storms; they are dependable as a source of dry firewood, have character and make me feel good. But they are largely dependent on Clark's nutcrackers to scatter their seed. Those curious mountain gossips and know-it-alls store the cones far and wide for winter food and actually remember the locations of most of those food caches. Some of what's left sprout as little whitebark pines, lucky for us. Whitebarks are princely trees.

Below them in the dark timber, azalea bushes were coming into leaf. The snowbanks were still deep, and rivulets of meltwater were running down the slopes. White marsh marigolds and globe flowers gleamed in the wetter places. The sound of water trickling, bubbling and dancing was constant.

We stopped higher up in the full alpine on harder ground by a particularly lush moisture-holding meadow, blanketed wild-potato white; the sun glimmered off the shining mountains across and down Tatlayoko Lake. Sage slowly fell to his knees and, using a pocket knife, began to dig corms out of the dark earth. He dug quietly, reverently, probably prayerfully, and placed his findings in a small bag for carrying. He wished to take a few "soont'ih" to give to elder friends of his up in Ulkatcho at Anahim Lake. "They haven't seen these for a while," he said.

Later we ambled along the ridge for another mile or two and stopped at a high point for more fabulous views and lunch. I sat, slowly munching my sandwich, a touch brainless, grokking it all in. I turned to my left to scan the eastern portion of the panorama. I could see across to Tullin Mountain above Chilko Lake, where Don and I walked in '98. There was faraway Ts'yl-os, dominating as always, and the wide expanse of big white mountains to the south of it. I spied a long, sloped, but roughly even grassy bench not far below me, with several vague parallel ruts, deep enough to be shadowed, running 200 yards or more along it. The noon sun was lighting up the slope. "That's it," I thought, after my slow double take and double check. "That has to be the racetrack." I sat silent for a minute or two to properly absorb this piece of news and then excitedly informed my friends of my surmise. It was farther north than I might have guessed, and it wasn't on the crest of the ridge in the picturesque fashion that I had fancied, but it seemed to me to be unmistakable. I could picture and just about hear the racers on their horses, hair and manes and tails flying, their relatives and friends cheering them on, making bets, everybody light-hearted, having a good time.

After some while, we ceased whatever it was we were doing — probably just sitting there in the warmth — and slowly made our way back down the ridge through the head-high krummholz toward the defile and the grassy flat behind, where the mountain trail had petered out. There were signs of old use: a tumbledown corral, weathered axe-cut stumps, a bit of barbed wire stapled to a tree, an old campsite, rusted cans half buried in the soopolallie brush, hints of old trail and more mats of spring beauty. I circumnavigated the two little swamp lakes, dark and still and near lifeless in the alpine fir; saw no signs of old-time use but for a small, rather wet-looking campsite; and rejoined my partners.

By this time, the sound barrage of resistant moo-cows was heralding the arrival of the herd in the lower wet meadows at the top of the steep trail. In the minutes it took for us to meet and walk around the cattle so we wouldn't inadvertently start a stampede back down the hill, they had pretty well destroyed the Indian potatoes growing there. The trail was a quagmire. The timing was lousy. I suppose the one, probably beleaguered, individual in the provincial Department of Forestry in Williams Lake, responsible for all grazing leases in the entire Cariboo–Chilcotin region, could not possibly have micro-managed this event, but a two-week delay might have allowed the meadows to dry up enough to better withstand the sidehill-pounding of a hundred heedless bovines.

So much for cattle in the alpine, but those golden ridges with the faraway blue snow mountains and the wine air are becoming addictive. My avocation and life practice of walking in high wild places and attempting to pay attention to where I am is a simple compulsion of sorts that must demand more dreaming and further exploration and understanding of the backcountry and the spirit of the old folks who travelled it.

Thus, it will come to pass that Peter and I will visit the Potato Mountains again in late September. We will contact Sage to see if he wants to come along, and we will walk that beautiful southern route. We'll climb easily up the steep, grassy slopes with the dead balsamroot sunflowers and the ribbons of poplar and willows, yellow and orange, the dry rills and rocky tarns, the clumps of whitebark and dark, purple-coned alpine fir, and eagles soaring. We'll walk up and over the rolling tundra and down to that pair of lakes, the "Echo" Lakes, where the storage pits, old camps and banks of massed spring beauty are. We will see the ancient, stone-lined steam pits. We will smell the smoke. We will imagine we hear distant voices and we will walk the ancient trails. The day will be endless and shining; the sky forever blue. The fine dust and wet-dirt trails will be detailed with animal tracks of every description. The early evening will turn golden and shadows will creep across the grass. The mountains will sit in their limitless glory, and the high glaciers will glint in the light of absolution. All will be still and all will be moving. Dogen's "Mountains and Rivers" will be "constantly walking," and Peter and I will begin to know our own walking. We will walk where the ancients have walked since "time immemorial," quite unaware that walks and trips and mountain expeditions cannot be repeated. You cannot push the same river twice. This will be of no consequence to us. We will walk with a light step, sure that we will be closer to knowing what the old days really were. You cannot know the old days, the old ways and the old-time people without knowing where they were.

Near the end of Sage's book, some of the elders quoted surmise that for Chiwid to make it through those long, cold winters at her little camps here and there across the Chilcotin, sleeping under snow, her feet hard as ice, eating her meat half-raw or frozen, sometimes howling, she was dreaming the spirit of a coyote. As she followed her seasonal rounds, walking back and forth across the land from one end to the other over all those years, they say she took on the soul and knowing of a coyote. One of Chiwid's

JS and Sage Birchwater above Tatlayoko Lake near the north end, looking out to the southwest. The hillside is covered in spring beauty and cinquefoil.

granddaughters said she used to say: "When all the coyotes died out, that's when I'm going to die."

But coyotes are still here, crossing the road in front of us in that unhurried, low-tail trot, pausing at the timber's edge to look over a shoulder to see what sorts of entities we are before fading off into the bush. Coyote lives. Chiwid lives.

Meanwhile, the dreaming continues unfinished. There are ridges and mountains to walk, the next valley to look down into, trails and camps and homesteads to seek out, and their inhabitants, settlers and aboriginals, dead and alive, to learn to know. In particular, William Wycott, old Stranger himself from over in that dry eastern Churn creek country, picks away at the edges of my curious mind and will not let go, and I find myself wondering as to his whereabouts and how he got there.

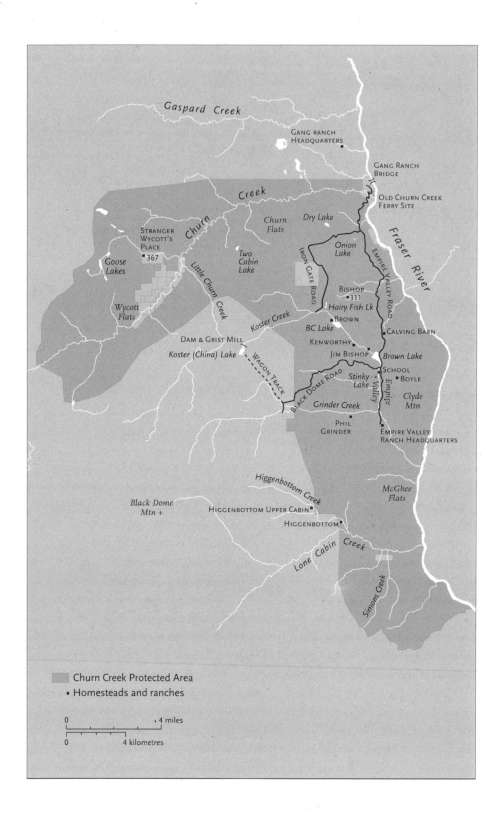

Gaspard Creek

Gang Ranch
Headquarters

Gang Ranch
Bridge

Old Churn Creek
Ferry Site

Churn Creek

Churn
Flats

Dry Lake

Fraser River

Stranger
Wycott's
Place
■ 367

Goose
Lakes

Churn

Two
Cabin
Lake

Onion
Lake

Iron Gate Road

Empire Valley Road

Little Churn Creek

Koster Creek

Bishop
■ 311

Hairy Fish Lk
■ Brown

Calving Barn

Wycott
Flats

BC Lake

Kenworthy

Brown Lake

Dam & Grist Mill

Koster (China) Lake

Wagon Track

Black Dome Road

Jim Bishop

Stinky
Lake

Empire Valley

School
■ Boyle

Clyde
Mtn

Grinder Creek

Phil
Grinder

Empire Valley
Ranch Headquarters

Higgenbottom Creek

McGhee
Flats

Black Dome
Mtn +

Higgenbottom Upper Cabin ■

Higgenbottom

Lone Cabin Creek

Simons Creek

Churn Creek Protected Area
■ Homesteads and ranches

0 — 4 miles
0 — 4 kilometres

SIX

Stranger Wycott's Place

Within and around the earth,
within and around the hills,
within and around the mountain,
your authority returns to you.

TEWA PRAYER

Stranger Wycott's place is just there on its bench above Wycott Flats up off Churn Creek on the west side of the Fraser River, mainly untouched, mostly unchanged, certainly not improved in any human way in a century, except for a forgotten galvanized-iron watering trough in some willow brush in a creek bottom. Gang Ranch cowboys have ridden through looking for cows; there is a line cabin up the hill a couple of miles away at the top end of one of the Goose Lakes. Hunters have been around there but seem to have stopped using the few old hunting camps in the area long ago. Prospectors and placer miners searching for gold up Churn Creek would no doubt have driven by and looked in when the track down the hill was dry and gold prices were up. But judging by the wary way yearling cattle respond when I've been down through there, I must have been the first person to actually walk in the area in a long time.

I don't know when I first heard of William Wycott, but from the beginning he came across to me as out of an old frontier dream. I read about him in Harry Marriott's classic *Cariboo Cowboy* and again in *Tales of the Ranches* in the People in Landscape Series, and by that time the pictures in my mind of this old character living up Churn Creek somewhere well past the back of beyond had formed themselves. Later, in the history appendix of the management plan for the Churn Creek Protected Area, I came upon a quick reference to a Wycott who lived on or by Wycott Flats and another to a Mariah Wycott who, given the context, must surely have been his daughter. The clincher for me was the brief suggestion that his homestead could still be found "back in the trees." Now that is a powerful, even mysterious, image, one hard to resist, of an old, half-hidden house, dark beneath poplars closing in, the place becoming inexorably more wild. I had to find the place, though I had only a vague idea where it was situated. How I did so is a whole other story.

All the while, the old homestead sits there in silence, sinking imperceptibly back into its sloped bench, a southeast and sun-facing open cove, out of the path of prevailing winds, between two small, converging creek gullies. The creeks are nearly waterless or are underground now in these warming times.

At the lower end of the slope above a willow and cottonwood thicket at the meeting of the creeks, the two-storey house stands in the open, tall and firm and near-black with time and sun. The first sighting through sparse timber — short, spindly poplars mostly — is imposing, monumental even, especially as the usual approach to the place is from downhill. The roof and floor are gone, but much else is intact, especially those beautiful, perfectly fitted corners. The house logs were broad-axed square and finely measured and fitted. Gaps between logs were painstakingly chinked with mud outside, and with thin, split poplar strips on the inside; there are few open chinks after a century of abandonment and exposure. Walls are thirteen to fourteen feet up, the door frame is made for a tall man, and there was no skimping on the height of the upstairs ceiling. The door and the few windows are, as was common for those times, framed without nails; all the securing was done with brace-and-bit-bored holes and hand-rounded doweling. But the ceiling joists are sagging and the floor joists are all gone; the careful notches to hold them at regular intervals along and between the bottom two side-wall logs remain sharp edged. The floor logs would almost certainly have been whipsawed flat by hand, by two hard-working people, one up one down, log by log. The ground around is rocky and dry and there is little rot, except on the low edge of the lowest logs on the northwest side.

William Walter Wycott's homestead up Churn Creek, pre-empted in 1884. The roofless two-storey building in the foreground is the house; the barn, with its shake roof still intact, is in the back. His irrigated field extended up the slope to the right.

On the sunny side there is an unusually large root-cellar hole. As often seems to occur with long-abandoned houses, spiny rose bushes grow where the floor used to be.

The barn is a short distance uphill from the house and is two storeys also. It stands with the same dark authority as the house. I'm guessing it's the older of the two buildings. Its roof purlins and nearly all its shakes, split from fir most likely, are still in place. The building is anchored by those same tightly fitted corners; the notch templates are identical, and the door is framed with the same hand-doweled thoroughness as the house. Manger and stall poles are built in and partially intact. There is a framed door out of the second storey on the uphill wall for hay, but the ceiling is gone and the joists neatly removed, for firewood I expect. The barn, too, has limited log rot, mainly on the back side down low. By the buildup of powdery cow dung within, it appears cattle still use this building for the purpose for which it was intended: shelter from the elements.

Below the buildings, close to the deeper gullied of the two creek beds, the one with a few stagnant pools farther down, are the remnants of a small apple orchard. One tree, healthily leaved — actually a cluster of tall stems out of the original stump, now gnarled and sunburnt — stands in tall grass. A half-dozen other such stem clusters stand nearby at regular intervals, but they're mostly dead and dry now and rattle when a breeze blows through them. There was bear shit in close vicinity to the tree and near standing timber the last time I was down there. This mid-Fraser region up off the river benches is not natural fruit tree country; that lone hold-out tree is a testament to the warmth of the cove where the homestead sits, and to the thought and care that went into the creation and maintenance of the orchard in the first instance. Think of the original cuttings, from Dog Creek maybe, or Ashcroft, wrapped and packed to preserve moisture and survive travelling, and carried in by packhorse sometime in the last decades of the nineteenth century.

The main room in the house up Churn Creek where the Wycott family lived approximately 100 to 120 years ago. This picture was taken in 2006.

There is a powerful quiet about this place.

Uphill above the barn, the clearing levels out somewhat. It is here where Wycott grew his crops, which in his era would have probably included wheat for bread. The slope had been carefully irrigated from top to bottom. Part of the upper end of the original field is in young fir now. A quick study of the understorey shows

no stumps, and the trees are of a uniform age and size. Wycott would have cleared the area, tree by tree, stump by stump, but it has grown back, likely sprouted all at once some warm damp spring when conditions for germination were just right and runoff water was trickling down the slopes behind. Without fire, grass-land edges creep in.

The dam site and dug-out flume track are bushed in but quite visible up the northernmost of the two creeks at a point where the hillside steepens and a head of water pressure can build. Relatively large ditches, all but filled in now after a hundred years of animal feet and general erosion, run down either side of the field. The one on the south side descends to above the orchard site. There are piles of picked stones along the ditch edges, one or two close to the house with old iron stove parts, rusted tin-can flakes and other metal bits on them, and more decayed debris scattered in the clos-est dry creek bed. Several smaller channels, still faintly visible in a slanting light in spring before new grass has begun to grow, run diagonally at intervals down the field off the main ditches. You need to feel their slight dip with your feet to be sure they are there. Finding them is like looking for the subtle grooves and depres-sions of old rock petroglyphs when the sun is high and the light is too bright.

I presume the kitchen garden was handy to the house, just on the downhill side, perhaps, where last year's grass stands in tall pale clumps in spring, or above the orchard where there are no stones. Partway along, off the south ditch, there's an area where barn weeds grow in a large, neat rectangle marking an enclosure, a chicken run or pigpen maybe. Rotted wood bits and crumbling saw-cut log ends indicate some kind of construction, probably a shed for shelter, along one side. Along the edge of the woods on the same side, a snake fence runs, still partially standing in a few places; most of the logs were massive, judging by the crumbled remains. There are similar signs of a large round corral with a gate entrance off it. The old-timers, it seems, built their corrals and fences skookum to contain their stock, most of which would have been half wild. Wycott could not have lifted those big logs by himself.

A short distance through the trees past where the gate would have been, a dark hillside rises with good, straight, mossy fir on it, many old stumps, and an obvious grown-in skid road links the slope to the cleared flat below. The source for Wycott's house and fence logs and firewood was close and uphill; Stranger Wycott was no fool.

If he had grown wheat, and many of the earlier pioneers did, he almost certainly would have gone over to Brown's grist mill, operating in those years at China Lake, closer to Empire Valley, to grind it. William Wycott and Samuel Lean-der Brown would have known each other from their earlier days as leading citizens over at Dog Creek across the Fraser. As the crow flies, the mill would not have been far away — maybe three and a half miles (six kilometres). He would have had to cross Churn Creek to reach it, but creeks run low in the fall season in this dry land. Churn, which rises on the northeastern edge of the South Chilcotin Mountains, has a large drainage area, and the crossing would be dangerous in flood. The 1:50,000 topographic map shows the route up Lit-tle Churn Creek, across to renamed Koster Lake, and past that down Grinder Creek to the Empire Valley Ranch buildings, as an almost straight line. An old land-registry map reveals a "Chil-cotin-Empire Valley Trail" running right through Wycott's place. I need to search around some future spring for signs of that trail or wagon track, particularly in that second-growth fir at the top of the sloped field or up that upper creek. The way in to the home-stead site from across the shallow creek course on the downhill side is quite wide enough for a wagon. On the far side of that dry creek are scant remains of an older, smaller building, more latterly used as a refuse pit: Wycott's original cabin, I think. It would be a prime site for old bottles.

Wycott had to settle for what was left

The details of Stranger Wycott's place are informative. They tell us that Wycott was an intelligent, skilled and resourceful man. As the log work on most remaining Cariboo road and ranch houses exemplify, many of the old-time house builders were capable log men, skilled with broad-axe, adze and draw knife, meticulous on detail. Stranger Wycott was as good as any of them. Those tight corners and squared, fitted house logs, the sheer size of his two-storey house and barn, the fact that they are still standing after more than a century, inform us he was ambitious and long range in his thinking, canny in his planning, and confident in what he was doing.

His choice of location is especially telling. He pre-empted lot 367 up Churn Creek on June 12, 1884. By that time, Thaddeus Harper, co-founder of the Gang Ranch, and other agrarian-minded, land-acquiring folks, some fresh from not striking it rich in the Cariboo goldrush, had beaten him to most of the best low-elevation land along the Fraser River. Wycott had to settle for what was left. Would-be ranchers in those early days looked for low, open land where the snowfall was minimal and their cattle could forage bunchgrass for themselves in winter. They tended not to go to the trouble of farming — that is to say planting, irrigating and mowing hay — at first, and cows do not paw through snow for winter feed like horses. Thus it was essential that the first pre-empters find available, accessible, low-elevation land of a suitable quality, with adequate water twelve months of the year, handy log and wood sources, and a southern exposure for maximum sun in winter and early spring. Winter snow melts late on those damp, sunless, north-facing slopes, and it is a blessing to have the home place out of prevailing winter winds. Wycott found all of these elements and more close at hand up off Churn Creek; he found a place where he could irrigate on a small and efficient scale for the kitchen needs of his family, and we know he had a family, a large one it turns out. That big root cellar by Wycott's house suggests many mouths to feed.

According to the indefatigable Clinton historian Don Logan, the BC census for 1881 indicates William Walter Wycott, born in 1836 and originally from Picton, Ontario, via California, was living at that time on the east side of the Fraser River, in the community of Dog Creek. He made his living as a pack-train operator between Lillooet and Barkerville during and after the goldrush years. As with other characters from the more distant colonial past, there are the usual uncertainties, confusions and differences of opinion regarding details of his life. In Wycott's case, some seem unsure about his first name, though everyone agrees on the nickname "Stranger"; it is a name that seems to fit. He got that particular moniker from his habit of calling most folks he met, especially ones he didn't know, "Stranger" or "Stranger man" in a deep voice. Otherwise, he is varyingly referred to as Tom, Carl or

William in some printed references, but mainly the latter; the first two names refer, more likely, to two of his sons. So I'll go with the majority, which includes Harry Marriott, who, from when he first worked at the Gang, appears to have known him quite well, and refers to him as William or even, in a light moment, "Bill."

There seem to have been no photographs of him, certainly none readily available. The lack of visual identification, his striking nickname and the general paucity of facts about Wycott add the beginnings of a mythic quality to our scant realization of him. The tidbits of information that come down to us are like snippets of faraway voices on a light breeze, the sounds fragmented, directionless, hard to hear and slightly ethereal.

Wycott pre-empted his first land claim, 160 acres down on the east shore of the Fraser River north of the mouth of Dog Creek, in 1868. He was appointed the first postmaster of Dog Creek in 1873 and ran the store there for a time also. By 1881 he was no longer postmaster, and, more seriously, he is listed as "widower." The woman to whom he was married in 1873, Annie Crang, from Bude, county Cornwall, England, had died two years later, in Clinton, at the tender age of twenty-three. Furthermore, she seems to have had an infant daughter, Eva Francis, born in 1874 but not listed in the 1881 census. Eva was issued a birth certificate, however, and William Wycott was definitely her father. In his sleuthing, Don Logan has found no other official mention of Eva's young presence, but there is an announcement of her wedding to Wiley Shermer in Ashcroft in 1896 and, interestingly, indications of descendants of the couple in various parts of the United States. It appears Eva was raised in the Lac La Hache area at 117 Mile, but we have no clue by whom. She probably had little or no contact with all the rest of the young Wycotts, her half-siblings older and younger, at Dog Creek and later up Churn Creek. We must wonder if, after her mother's death, she ever met or saw her father.

That same 1881 census indicates that Wycott already had several older children, and baptism records indicate the mother of the three oldest as Matthilda "Maggie" Kwonsenak, an aboriginal woman from Canoe Creek to the south up off the Fraser River. Those three Wycott children that we are most certain about are Tom, the oldest, aged twelve in 1881; George, age ten; and Mariah, age eight, who seems to have been born early in 1873, the same

year William and Annie were married. Also listed is the mysterious "MJ" (Mary Jane?), who may have been five in 1881, whose
mother is not named and who we do not hear about again. As well,
there seem to have been two other boys, of indeterminate age,
named William "Whitecott" and Johnny Wycott, whose mothers
are listed as "Kwintchinik" and "Conchenak,"
respectively. Those names sound so similar
— surely it was Maggie Kwonsenak herself who
was their mother. But are we certain William
was their birth father? The proximity of Annie's
premature and inopportune death to the prob-
able birthdate of "MJ" raises the possibility that
she died during childbirth, an all-too-common
occurrence in those times. Mariah would have been just two then,
and we presume she and her various brothers had been looked
after by their mother Maggie all the while.

He was appointed the first postmaster of Dog Creek in 1873

Life is seldom simple, and it was obviously not so for Mr. Wycott.
There are many mainly unanswerable questions here, but one fact
stands out. Much of his life story centered around his two partners: the short-lived and briefly married Annie Crang from back
in the "old country"; and Maggie Kwonsenak of Canoe Creek and
her offspring.

Who was Annie, and how did she come to join William in the
new world? What drove her to cross the wide Atlantic Ocean and
this vast North American continent and finish up out there in that
beautiful backwater country around Dog Creek? How did one so
young and such a long way from her homeland come to die so
soon, in the bush, of, as stated in the death certificate, "congestion
of the brain," whatever that might mean? Was she a large woman?
Small? Pretty? How would she and Stranger have met? Could it
be they laughed together? Was he kind to her or harsh? Was her
union with him liberating? Or simply servitude?

Similar questions can be asked about Maggie, William Wycott's
first partner and the mother of Thomas, Mariah and the rest.
What sort of person was she? Was it her young beauty and charm
that attracted him or merely her youth, strength and the fact that
she was female? Could their relationship actually have been — at
the beginning, at least — an affair of the heart? Did they have a
true partnership, as long as it lasted? Or was her relationship with

William simply an arrangement of convenience?

For sure, William Wycott dropped Maggie and the older children fast when English Annie appeared, a quick and probably typical termination of a "country marriage" of that era. What could Maggie have felt and thought when she learned, like a myriad of other Indian women in those times, that she was being replaced? Or was this just standard frontier business? Where did she and their children go? Back to her family at Canoe Creek we would suppose. At the very least, Wycott would have lost a workmate and a companion, not to mention the mother of his children. Did he gain a similar partner in Annie?

Two years later, after Annie's death, William picked up his old arrangement with Matthilda Maggie Kwonsenak and the children almost as quickly as he had abandoned them. Maggie must surely have been a forbearing and devoted woman. Or perhaps she was simply pragmatic? Was she open-hearted? Willing? Happy? Or desperate and resentful? How did her life pass? Was Wycott good to her? We can well imagine she was strong and hardworking, and we know she raised between five and nine children, depending on how many survived their young childhood.

Veera Bonner, in a brief entry in *Chilcotin: Preserving Pioneer Memories*, mentions a William Wycotte (note the "e") who had a place on Churn Creek near Wycott Flats, obviously our man "Stranger." After citing June 1868 as the date of the Dog Creek pre-emption, she makes a terse reference to two additional sons, Fred and Cal. Don Logan in his census searches confirms their existence and that they are younger than the original brood, and he adds the name of another son, Nelson, younger yet. Maggie is the mother of each.

William Wycott was one of those first-generation colonists, who came to the Cariboo goldrush to get rich or make a living one way or another, for whom there were virtually no women of European descent available to marry. The presence of Annie Crang would have been an exception to the usual rule. More typically, Wycott and other settlers of his area and era looked for wives and companions from among aboriginal women of the local Shuswap (Secwepemc) bands at Canoe or Dog Creek, or Alkali Lake. (Wycott's original pre-emption, which he maintained for sixteen years, at least, was just down the hill on the Fraser River from the latter two places.) Such marriages were common practice for those earliest settlers like Herman Bowe, Joseph Haller, Phil Grinder, Conrad Kostering and Magnus Meason, to mention examples of well-known early Cariboo pioneers and family men. What else to do? Such arrangements seem frequently to have been of mutual benefit. Phil Grinder's wife of long duration, Nancy Kala'llst, for example, a member of an old Gaspard Creek family, came from nearby, either Alkali Lake or downriver at High Bar — sources differ. The old couple, founders of generations of Grinders to come, was "respected and liked by all," observed Harry Marriott. The Cariboo–Chilcotin region is peopled with the descendants, acknowledged and unacknowledged, of such unions today.

> *The sounds of children playing would ring out in that clearing*

In 1884, nine years after he rejoined his family, William Wycott claimed his second pre-emption, lot 367 up off Churn Creek on the Chilcotin side of the Fraser River. He had a renewed partnership with Maggie and an expanding, double-barrelled brood to help keep alive and healthy, which would later in their lives, hopefully, when the elders slowed up, return those efforts. That year or shortly after, the Wycott household crossed the big river and made its way westward to the long, bighorn, grassland benches and breaks we now call Wycott Flats and the open, southeast-facing basin above them on the northwest side of the valley where the old house and barn stand today.

The sounds of children playing would ring out in that clearing up there above Churn Creek in those years, and the sound of their mother calling them, perhaps, echoing on a cooling autumn evening when sounds carry and reverberate, animals call, the trees go dark, dew falls and the pale smoke of the supper fire hangs low. Were there good times? Loving moments? Was there injury? Illness? Deprivation? Heartache? Booze? How many of those children endured to their adult years? We hear no further news of George, the second boy of the original family, or of the little girl, MJ. What became of them? Was there favouritism? Neglect? Anger? Abuse? Those small survivors would grow up tough and familiar with a life of hard labouring. Where did they all go when they fanned out into the big wide world? We can assume the boys went out to work on local ranches. Veera Bonner states with certainty that Fred and Cal did so. Would the daughters, country-wise to be sure, become good homesteader wives? Most likely, "yes" would be the answer.

The three youngest Wycott sons are listed in the census of 1898 as "labourers," which translates as "cowboys," at St. Mary's Creek, an early name for what is now Churn Creek, above which the Wycott home place is located. William would have had several strong young backs to lift those house and corral logs.

Thomas, the oldest from the first cluster of Wycott offspring, was twenty-nine years of age in 1898 and is differentiated as "farmer," which implies he owned land. It turns out that when Wycott registered his 1884 Churn Creek lot, he also claimed, in young Tom's name, an additional piece of what I suspect, from a quick look at the topo maps, to be summer grazing land. This property, lot 366, small, open, swampy and probably drainable, was on the main branch of Gaspard Creek only a few miles away over a low divide. Tom was then fifteen.

William and Matthilda Maggie Kwonsenak were formally married much later, on March 4, 1903, about a decade before his death. He would have been in his late sixties; she was younger. The Roman Catholic Church had a growing influence among Indian peoples of the central Interior in the late nineteenth century. I have to wonder if there was pressure, direct or otherwise, to make an "honest woman" of Maggie, or was it simply that William was feeling his years, grateful for her presence and willing to formalize the bond. Or were they officially conjoined for some other reason altogether?

The day after my first walk down there, I asked Chilco Choate over at Gaspard Lake why the Wycott family settled way up Churn Creek like that. Choate knows the Gang Ranch–Churn Creek country as well as anybody, having lived and hunted in the area for virtually all his long adult life.

His answer was simple: "bunchgrass." Bluebunch wheatgrass is the most nutritious feed for stock, it maintains its food values when frozen, and the open south-facing slopes and flats along Churn Creek had low snowfall and good conditions for it. The winds along the main valley blow much of the snow away. That's an important reason why bighorn sheep winter along there. Cattle don't survive well in most places without hay, but they have a better chance of wintering in a place like Churn Creek. All the old ranches lost cattle in bad winters in the old days; the Gang Ranch was famous for it.

I extolled the intelligence of the situation and layout of Wycott's place, using it as an example of how smart I thought the first settlers needed to be to wrest a living out of remote bush. Chilco said with some energy that not all of them were smart. He thought quite a few of them were "damned stupid," in fact. He listed the qualities of some of the places he'd seen: bad water, lousy feed, no handy wood supply, dark in winter. It occurred to me that nature would have its ways of taking care of the unready.

The location of William Wycott's original (1868) pre-emption is exactly known, but its function is predictably ambiguous; hard facts on matters of early history are usually scarce. To add to the ambiguity, there are, in fact, two flats in the Churn Creek–Fraser River area ascribed to him: Wycott Flats, nine or ten miles up Churn Creek below the 1884 homestead and now part of a protected area for California bighorn sheep and for the bunchgrass upon which they so depend in bad winters, and his original pre-emption, Wycott's Flat, along the east side of the Fraser below the mouth of Little Dog Creek, which is now a piece of Indian reserve land. When Stranger Wycott made his living as a pack-train operator, he would have needed low-elevation wintering grounds for his pack animals close to his main business route on the east side of the river. Furthermore, the major early trail to the Cariboo goldfields ran along the river through his pre-emption. Possibly he had dreams of running a roadhouse or store, like so many of his ilk. A copy of an old Royal Engineers map from 1861 covering that mid-stretch of the Fraser shows

Nature would have its ways of taking care of the unready

several "trading posts" up and down both sides, but especially the east side, of the river. Wycott's Flat on the Fraser River would have filled the bill, though it would have been a hot, airless and unforgiving place for a settler down there in summer. Most ranch headquarters and Indian communities are off the river and uphill, typically at an elevation where the dry, sagebrush grassland slopes and the semi-wooded parklands come together and breezes blow.

In the early 1880s, up and down the Fraser River, Indian reserves were being established. I note that in July 1881, Wycott's Flat was purchased by the Reserve Commission, "intended as a fishing station" for the Alkali Lake Indian band. I conclude that William Wycott could have used the ready cash and was glad of the opportunity to sell. Perhaps he wanted to get away from Dog Creek. He would need the money to finance his new pre-emption and the establishment of the ranch buildings up Churn Creek, not to mention the perennial maintenance of that growing clutch of young Wycotts.

"How did he get in and out of his place?" I asked Chilco Choate. I had seen that the country was gullied and steep along and above Churn Creek and up in behind his homestead, and there were no obvious signs of a wagon track out of there. Chilco replied that Wycott went in and out with packhorses. He parked a one-horse cart in behind Bear Springs somewhere and drove the rest of the way down to the Gang Ranch store or over to Dog Creek for supplies. That trail over the hump piques my curiosity; I'd like to take the time to circle around and find that route. Does it run straight northeast through the rocks and clay banks above his place, as Choate implies, or did he take the Chilcotin–Empire Valley Trail farther to the northwest, veering down Little Gaspard Creek through Augustine's and William's Meadows and linking up with the main Gaspard Creek drainage and trail to the Gang Ranch headquarters? Or were both routes valid?

> *He has become*
> *one of those*
> *semi-legends*

William "Stranger" Wycott seems always to have been an interesting character, but if we go by the stories that come down to us, he got downright eccentric in his older years, due in part, no doubt, to the isolation of his life up Churn Creek. He has become one of those semi-legends in the Cariboo–Chilcotin you hear or read about, another of those examples where fragments of history merge into myth, something like the Chilcotin War and that row of graves up at Graveyard Valley near the headwaters of Big Creek; Chiwid (The Chickadee); George Turner (on the run from the Dalton gang down in the States, supposedly); Thomas (Domas) Squinas, old-time Ulkatcho chief, and the great hunter-packer Josephine Robson from around Anahim Lake (daughter of trader Anton "Old" Capoose), both renowned in their time for their exceptional woods, horse and survival skills, and, in the case of Squinas, for his leadership; grizzly bears and wild horses in the backcountry; and the way stories about Frank Chendi ("Jack Pine"), who walked large tracts of the Chilcotin wilderness barefoot for decades, so they say, and who lived under the most minimalist and demanding circumstances with his goats down by Chilko Lake, are spreading around the region now. You start talking about old-timers and old ways with folks, ask a question or two, and somebody, sooner or later, brings up Stranger Wycott.

People wonder how he lived up there all those years by those flats named after him in that Churn Creek area somewhere. They talk about his toughness, his lice (his "galloping eczema," as Marriott puts it), his thieving, and those definitely legendary giant long-horned steers and bulls of his that got away on him in his last years and grew wilder than deer, five times as big and dangerous, and just as fast. After he died, it required major efforts to chase them down, round them up and drive them to the railroad at Ashcroft to sell. Some of the too-tough ones were shot; if left to live, they were a disruptive influence on more domesticated stock.

Even before he got elderly, Wycott chose not to sell his steers until they were eight years old; he wanted them to gain their maximum weight, I suppose, and I understand they drive and survive better when they are mature. Judge Henry Castellou's account of Wycott pushing his herd of oversized beasts down the main street of Barkerville is most entertaining. In the midst of rains and runoff when the water table was high and Williams Creek was overflowing, the street was a sea of mud and running water and it was necessary to drive the cattle through all of that to reach Stouts Gulch, where they were butchered. The arrival of Wycott and his wild steers, some with horns three and four feet long, was an event, and the population of Barkerville would turn out for it. People barred the doors and retreated to second-storey viewing points to watch the show. When one of those animals barged through a saloon door and into the barroom, as apparently happened, the resultant mayhem, redistributed mud and wreckage was stupendous.

Then there are the stories of his exchanges with various representatives of the "civilized" world. Like the time some society lady visiting Gang Ranch from the big smoke down on the coast rattled on and on about how important and sophisticated she and her accountant husband were and how she was related to Queen Victoria. Wycott, who was known to normally wear buckskins, was present, dressed in his best clothes for the occasion. So old Stranger is standing there, with his great, tobacco-streaked beard, his long flowing white hair, his two straw hats — one new and one old — his way-too-small suit pants and tight dress jacket and the rope for a belt, his laces-less shoes and his general murkiness, and he's looking hard at this self-styled dignitary as she carries on. When she's finished, he stares her straight in the eye and asks, in that deep voice of his, "Stranger lady, have you ever had the piles?" He rode a lot, and we have to believe his own had been acting up for a while.

In his latter years, old Stranger's reputation as an inveterate thief increased, and the stories are rich. Hilary Place tells how Andy Stobie, foreman of the Gang in the early years of the twentieth century, having just bought a new workhorse harness and hearing that Stranger was on his way down to play poker, guessed the old reprobate would be gunning for the brand new tack when he saw it. Apparently, Wycott travelled almost entirely at night. He unhitched his team and hung his old harness on the first peg in the barn; the new Gang Ranch gear was on the second. During the card game, Stobie snuck out and switched the two harnesses. Later that night, Wycott, feeling around in the dark, wound up making off with his own harness.

Another time Stobie spied Wycott stealing a pound of butter from the Gang Ranch store and hiding it under his hat. It was winter and the fire was on. Stobie didn't say much but shuddered and commented how cold he was. He cranked the stove up good and hot and, despite Wycott's efforts to get away, kept the conversation and firewood rolling until melted butter was running down old Stranger's face and neck. "That's an awful brand of hair oil you're using there, Stranger," Andy Stobie is supposed to have commented.

William Walter Wycott died in 1914 in the spring of the year, according to Harry Marriott. Others say earlier. He died at the Gang, having made some kind of arrangement to sell them his place and live out his last years there in a little old cabin behind the blacksmith's shop, a classic, old-fashioned, Chilcotin-style "retirement plan." The story has it that he had nine $100 bills sewn into his (probably less than freshly laundered and ironed) long underwear. They say his son Fred, who I assume to be the oldest of the last set of Wycott boys and who cowboyed for the Gang, inherited the money from the sale, a considerable sum. He went out on a year-long toot, re-emerging at the Gang ranch cookhouse on the following Christmas Day, "drunker than a hoot-owl" (however drunk that is), totally broke and quite happy to resume the old cowboy life. It seems the responsibility of packing that much money had weighed heavily on him.

The fact of Fred, the third youngest of all the Wycott children, as inheritor raises a further question: where do Thomas the "farmer," George, if he survived, or William and Johnny, not to mention the women, Mariah and the unknown MJ, fit in here? Why did they not qualify as heirs? Did they disappear? Were they dead? After due deliberation I have to ask: did William marry the steadfast Maggie to ensure a rightfully legal heir to the Wycott estate? And the rest of the offspring, the disinherited, where did they all go?

Mr. Wycott has left his mark, not only on the landscape up Churn Creek and the pair of flats named after him and in the stories that have come down to us, but also through those descendants who bear his name and live around the country to this day, most specifically at Alkali Lake and around Williams Lake. His

Annie (Higgenbottom) Grinder and Mariah (Wycott) Higgenbottom after a successful hunt, circa 1928. Annie is carrying the rifle; Mariah, on the right, is the mother of Annie.

Eva Francis (Wycott) Shermer of Chehalis, Washington, was the only offspring of William and Annie (Crang) Wycott. She was the half-sister of Mariah and was born within a year of her. This photo comes to us from Eva's granddaughter Bette Gaerisch of Aloha, Oregon. Stranger Wycott was her great-grandfather. Thanks to Don Logan, Bette has recently learned of BC relatives on the Maggie Kwonsenak side of the family that she never knew she had.

oldest daughter, Mariah, was likely a major source of care to the younger offspring and a pivotal person among the Wycott children, I would guess. Don Logan reminds us that she and Annie Wycott's daughter Eva Francis, the only Wycott child with a non-aboriginal mother, were born within a year of each other. They would have led parallel but sharply contrasting lives. However, it is Mariah, raised in familiar country at Dog Creek and in the bush on that sunny hillside up off Churn Creek, who is of particular interest to me. How could she have kept an open heart through the changes and stresses of her young life? Or was it that she simply grew up hard, a victim, a survivor?

Mariah, "related to the Churn Creek Wycotts," as the history appendix in the Churn Creek Protected Area management plan defines her, came eventually to live common-law with Harry Higgenbottom. He was a horse herder and the original pioneer at

Higgenbottom Creek up and off Lone Cabin Creek, a deep crack in the landscape south of Empire Valley Ranch, around the turn of the twentieth century. I've looked over that country down there from the north side and, again, from up on top of Red Mountain to the southwest; it is starkly beautiful but rugged and remote, and it is hard to imagine living in it. The location of their place at the bottom of a steep basin in the depths of the Lone Cabin defile, surely sunless and cold in winter, makes Wycott's homestead seem positively accessible and light-filled in comparison. Mariah and Harry were to begin several generations of Higgenbottoms in and around the Clinton area.

The photos of Mariah and Annie Higgenbottom and of Eva Shermer are instructive. Mariah's family remained up there in the mid-Fraser River region of central BC and carried on in the rancher-hunter, survivalist mode. Eva and Wiley Shermer moved around and spent time in Atlin in northern BC, where they mined gold, but later settled into farming down at Chehalis in Washington State. Note the orchard setting in the picture (circa mid-1920s) and the transformation of the automobile into a farm tractor. There is the whiff of mythology here, a sense of the classic, ages-old divergence of and struggle between hunters and farmers, Abel and Cain.

The lives of many wives and children of those frontier relationships, especially in the backcountry, must often have been difficult — for some even hellish. In his brief history of the Empire Valley Ranch, Henry Koster, whose parents had recently taken over part ownership of the place, describes an episode involving their neighbours Harry Higgenbottom, Mariah and her children. Louise Garrigan, a daughter of Mariah from an earlier relationship, arrived at the Koster home "all beat up" and worried for the safety of herself and her (I presume younger) brother Henry Garrigan. She said they were being physically abused by Higgenbottom and that he was making them kill Empire Valley Ranch calves, butcher them and bury the hides. Higgenbottom was reputed to bully his family, and Koster's mother and father (Henry Koster Sr.) found Louise's story believable. After she showed police where the hides and heads were buried, Higgenbottom was arrested, convicted of child abuse and cattle rustling, and sent to jail in the Lower Mainland, where he was killed in a

fight. Mariah and her maturing family remained neighbours of
the Koster family for years.

Coincidentally, in Don Logan's book on the Big Bar Mountain
pioneers from down across the Fraser River, there is a photo of
Louise Garrigan with Annie Grinder (nee Higgenbottom), Henry
Grinder's wife. Both young women, presumably friends, are
dressed in their best and holding cats in their arms; a dog sits at
their feet with that proprietary look most dogs get when they're
with their own folks. The picture looks as if it was taken on the
open slopes of Big Bar Mountain, where the Grinders lived. Both
women are unsmiling, determined looking, serious. Louise is
blond haired, Annie dark. I have studied Louise's hewed face a
little, tried to peer into her shaded, distant eyes, and wondered at

*Annie Grinder
and Louise
Garrigan,
likely at Big
Bar Mountain
or Big Bar
Creek, circa
1930s. Annie
was married
to Henry
Grinder; her
father was
Harry Higgen-
bottom.*

her story. Annie Grinder was her half-sister; Mariah Higgenbot-tom (nee Wycott) was mother to them both. Did the Grinder fam-ily take her in? Did she find safety? Where did she live her adult life? Was she loved? Or used? And Garrigan, her father and her mother's first partner, who was he? Why did he eventually disap-pear from the family scene? Did he die? Did he get tired of it all and leave?

Branwen Patenaude notes in *Trails to Gold* that a Phil Garri-gan and his son Pete, both blacksmiths, who lived near the foot of the original Cariboo Road over Pavilion/Carson Mountain north of Lillooet, ran a roadhouse and freighting business there in the late nineteenth century; in other words, they got around. The son would have been roughly the same generation as Mariah. Though travelling was a slow business, south-central BC was a small place a century ago. Interestingly, there is an old burn on the remote south side of Black Dome Mountain, the side overlooking the Lone Cabin Creek gorge, known as the "Garrigan burn." It was probably intended to create summer graze. Because this is long-time Empire Valley summer range, Pete Garrigan must have worked for Empire Valley Ranch at one time.

And that tiny ragamuffin Florence Garrigan, who shows up in a picture of a group of Jesmond school kids, Grinders and Pigeons mostly, in Don Logan's Big Bar Mountain book — she's the blond one holding the doll — is she Louise's daughter? If so, Stranger would have been her great-grandfather.

So what were the options for Mariah, and for her children? How was her own young life, her growing-up years at Dog Creek, and her later years out there in the back of nowhere above Wycott Flats? She would have been about eleven years of age when the family moved across the river. Was she loved, appreciated? Or ignored, just another body to help maintain the home place? Was her father a warm or a cruel man, or both? Such two-edgedness is possible. Did her laughter ring out across the clearing on those cool autumn evenings with the sun dropping down and the blue smoke rising? Did she have dreams and plans? Did she learn to read? Was her young heart capable of being touched by small events, personal moments, acts of kindness? Or was there sadness and despair? And all along, hcr brothers, bigger or younger, did they protect her? Or abuse her? What choices did Mariah not have?

As a young female, she would be valued. Young horsemen for miles around would have come prowling; older cowmen would have come looking for a house mate. Women, in those times, especially daughters of pioneers, hardwired to a tough life, were in demand as companions, workers, wives, bearers of offspring. Did she worry about the future of her own children, have hopes for them? Or did she live in fear and distrust? Based on his reputation, it seems quite possible that Higgenbottom mistreated Mariah. What was there to stop him, down there in that dark hole off Lone Cabin Creek? Surely she would have seen that he was abusing her Garrigan children, and possibly her offspring with him also. Most mothers feel and know such things, though some don't, and some deny. How could such treatment, such fear and pain and helplessness, such anger, be hidden in a society of just one family? Did she need to fight to defend her brood and herself; did she stand up? Or was she resigned? Silent? For survival, did she aid and abet and bury the hurt? Abuse begets abuse.

Henry Koster states that after Harry Higgenbottom's departure, Mariah and her clan were "good neighbours." Higgenbottom's horse herds were rounded up and probably culled and sold; their small herd of cattle ran free of charge on Empire Valley land. The implication would be that, to some degree, peace returned to Higgenbottom Valley.

I have long held a belief that on the frontier edges across the West, abuse — that is the expressed power of the physically strong over the powerless: women, young children, the aged and slow-witted, the very poor — was as common as dishwater. Aside from personal ethics and self-management, there would be little in the way of social sanctions to stop it for reasons of isolation, lack of accountability and desperation, and because life was hard. Abuse helped us cross this vast continent and get here. We rightly acknowledge and honour our courageous pioneer ancestors for surviving and for settling our western country and doing the very best they could, but abuse of the weak and exploitable, whether human, animal or wilderness itself, is central to our being here, all the more so because it is not normally recognized as such. There is unfinished evolutionary business in front of us here. Abuse, in its various expressions, past and present, festers our collective soul.

And what about the rest of the Wycott progeny? What kind of

folks were they? Where did they wind up? Did some of the boys cowboy over at Alkali Lake Ranch or at other ranches in the region? Did they become part of the social world around there, settler or Indian, or some hybrid world between? Did Wycotts beget more Wycotts? The answer would be a general but unconditional "yes"; there have been Wycottes in the First Nations community of Alkali Lake (now named Esketemc) and in Sugarcane by Williams Lake these years, and they now spell their surname with an "e." Sage Birchwater, who lives at "the Lake" and who knows many of these folks, tells me that's just language evolving, and he's right, of course.

Abuse

begets abuse

In the summer of 1964, my partner Marne, a teenager then, worked at the Chimney Lake place, a guest ranch. While there Marne developed medical issues sufficiently serious to spend extended time in the Williams Lake hospital. Her roommate was a cheerful, diminutive, middle-aged Indian lady from Alkali Lake, in hospital for blood poisoning, who referred to herself most definitely as Mrs. Pascal (Annie) Wycotte, with an "e." The man she married was likely a great-grandchild of old Stranger. The two roommates had a right good time, talking, laughing; Marne showed Annie Wycotte origami, to the latter's great amusement, and Annie told Marne stories of her family, including one involving "sniks," which turned out later, to Marne's relief, to translate as "snakes." Annie taught Marne to say (approximately) "la-am pos tshene," which, translated from the Shuswap (Secwepemc) language, means "my heart is happy," as Marne remembers it. Annie read Marne well. Mrs. Wycotte spoke with pride of her sons, who were working on the hydro lines at that time. She and Marne continued to correspond for a while. On a Christmas card Annie sent, with Santa's sleigh and a string of reindeer on the front, she drew an arrow pointed at the deer, with the words, in neat blue ballpoint, "good to eat."

And how were those later aging years for William and the faithful Maggie from down river, the dutiful one, the one who raised their children? Did she, like so many, contract TB or influenza or some other kind of early bacterial death? Could she live out her full "three score years and ten" with him, or did his latter-day eccentricities finally wear her down? In the end, was she driven back across the Fraser River, to Alkali Lake maybe, or back down to her parents' people?

From the stories that come down to us of those last years, when Wycott was losing his edge, it sounds very much like he was on his own up there above Wycott Flats: alone after his big steers and his sons got away from him, before his horses went wild and his lice ganged up on him, before his old dress clothes got tight and worn through and society lost much of its meaning for him, before a marginal diet and the ailments — the piles, bad back, arthritis and pulled and broken teeth — caught up to him, before he got bushed and moved down to the Gang. Perhaps his greatest loss and greatest need all through was companionship. My intuition says bachelor Stranger was alone up there for a very long time.

This William Walter Wycott, traveller of the night, this old man winding down and wearing out in that backcountry up there behind Gang Ranch, I wonder, did he still laugh? Could his heart still stir with hope in the spring of the year? Could he hear that haunt in far-off sandhill cranes? Because I have this low feeling that the trees were leaning in over him on those last long black nights of winter, that the gravel was falling and rattling in the rocks up in behind him, up there between those two small creeks. What was it he heard whispering as he rode home in the dark? Did living become nothing but necessity, day by day by day? Had he become just some cantankerous old individualist existing, too ripe and ornery for regular folks to connect with? Had he grown fearful as his old home worlds disappeared? Was that his own voice he cussed at as he prowled about his clearing? That harsh Chilcotin land, did its vastness close in over him in the end?

Could it be that the story of Stranger Wycott takes on the patina of myth?

Long after his departure from this good earth, Stranger Wycott's place carries on out there in wildness, settling slowly into that piece of hard, mostly cold, ground, settling into silence, stillness and the deeper recesses of time, where time itself turns timeless. Wind and rain and sun and freezing and the feet and mouths and offal of passing animals affect it, but the effect is minimal. The outcome is a settling of the subtle aspects of the place, a slow determining of what will be where, a delicate flowing and gentle filling-in of spaces, a slow changing and turning and coming to balance. At the centre of balance is stillness. Wild, like myth, knows to stand still.

The impact of events and time at Wycott's place, and across the land, may be light, like dust, but even one call from a flying bird passing one time leaves a mark. The buzzing of bees, the mosquitoes humming, the low chuckling of birds in the underbrush, the trees' movement and murmur lay a lucent blanket of sound around the clearing that we cut through finely as we walk. We become watchful here, leaving only a gentle swirl of motion as we move. As we watch we come to sense the world is watching too, and we are joined in our watching; awareness becomes aware of itself. When we are light, we are not deeply differentiated nor separate, but merely one more element in wild nature. In light, we hear the crackling and roaring of June grasses growing at full solstice if we put our ears close enough to the ground. The summer sun infuses all it reaches with its light and energy, and leaves, grasses, flower husks, from their depths, reflect that accumulated deep light back slowly, as the sun's angles lengthen and fall colours glow. Stranger Wycott's place is more wild now than before he settled there, more wild than those last elder years of his, alone, falling back into himself. And we are a little more wild for having heard his story.

When you find your place where you are, practice occurs.

                                          DOGEN

ACKNOWLEDGMENTS & PHOTO CREDITS

For their help, direct and indirect, in the creation of *Stranger Wycott's Place*, I am truly grateful to the following: Sage Birchwater of Williams Lake, who helped to get me going by his example and his encouragement; the always helpful Dennis and Iris Redford of Driftwood Camp at Tatlayoko Lake; Veera Bonner of Fletcher Lake near Big Creek, to whom I dedicate "In the Details"; Hennie and Ria Van Der Klis of Chilcotin Lodge at Riske Creek; Chilco Choate of Gaspard Lake; Mike Brundage of the Clinton Museum; the indefatigable Don Logan of Clinton for all his help and sharing, and Karen Logan for her patient support of Don's history and old bottle addiction; Van Andruss at Moha in the Yalakom Valley; Kenny and Joy Schilling of Darfield up the North Thompson River for having helped maintain our connection since 1947; Barry, Warren and Casie Menhinick of Goldbridge; Russ and Carol Dreger of Knutsford, south of Kamloops; Gerry and Sharon Betts of Port Alice for the clams and their long-time friendship; Jim and Anne Sutherland, good old friends from some place east of the Rockies — Tranna, I think it was; Bette Gaerisch of Aloha (Portland), Oregon; Gary Snyder of Kitkitdizzi, Nevada City, California, for his exemplary practice, graciousness and support; Don Brooks and Dana Devine of UBC for the walks, the birding and annual backpacking trips, the horse expeditions, the incredible cooking and all that good wine; Louise (Betts) Smith from Vancouver, but originally from Port McNeill in the good old days, for her spirited encouragement; and Bob and Mary Steele of Point Grey, the best of the best, for having parented the best (Marne), and for their great friendship and support.

Closer to home here in Victoria, I am grateful beyond measure for the friendship and interest and frequent walking trips, short and long, to Peter Stein, Trevor Calkins, Bob Whittet and Tom Hueston. Thanks also to my patient computer guru, Derek Fuller next door, and to my brothers Chris and Andrew, who love this land as deeply as I and who go back to the beginning. My gratitude goes as well to the good folks at New Star Books, in particular Audrey McClellan, ace editor, and the generous Terry Glavin, who read *Stranger*, understood it and liked it, it seems, and decided it might be good enough to publish and spread around. And Peter, thanks for the book title concept.

A special acknowledgement to Padmananda and Tara Padmananda, my Yoga teachers for decades, to you "in light"; to my parents for kicking off this particular go-round and seeing the first part through with care and for answering all those questions; and above all to Marne and her uncompromising love, intelligence and support. And thank you, muses, palpable presences, whoever or whatever you are.

On behalf of Stranger Wycott, Harry Marriott and all the other voices from the old days and the back country, here's to all of us who love these places and here's to this living land that supports us.

I had a fine collection of photographs to choose from for *Stranger Wycott*; the quality of those chosen speaks for itself. Marne St. Claire contributed the greatest number, taken on our trips together over the years, and my brother Chris kindly accompanied me on two tours recently, expressly to take some fine pictures for this tome. Tom Hueston, Peter Stein and Bob Whittet, all lovers of the wild and my companions on various trips, are responsible for most of the rest of this current set of photos. And a very special thanks to Don Logan, who generously made copies of some wonderful historic photocopies in his personal collection for me to use. I dedicate the Stranger Wycott story to Don; I could not have written it without him. Thanks also to Bette Gaerisch for kindly providing photos of Eva (Wycott) Shermer. Marne is responsible for the cover picture. My deep gratitude goes to all you picture takers for your friendship and contributions. Your efforts will help these stories live.

I am informed by a great many sources; Leslie Marmon Silko, a novelist and poet from Laguna Pueblo in New Mexico, is one of them. She wrote a fine book called *Ceremony* that emanates from another, more southerly, part of this same wild, western North American land. She introduces her story with a short set of verses about Thought-Woman the spider, weaver of webs.

Whatever Thought-Woman thinks, happens: she thinks of her two sisters and they appear; the three sisters together create the universe, which includes this world and four other worlds below it. As she names things, they appear. She is sitting in her room now, thinking the story that is *Ceremony* as the author transcribes it for the reader's edification.

My question to you is this: who or what is creating these Chilcotin stories in this small collection that appears before you now?

RESOURCES & FURTHER READING

ANTHROPOLOGY / MYTHOLOGY

Bierhorst, John. *The Mythology of North America*. New York: William Morrow, 1985.

Bringhurst, Robert. *A Story as Sharp as a Knife*. Vancouver: Douglas and McIntyre, 1999.

Carlson, Keith Thor, ed. *A Sto:lo-Coast Salish Historical Atlas*. Vancouver: Douglas and McIntyre; Seattle: University of Washington Press; Chilliwack, BC: Sto:lo Heritage Trust, 2001.

Corner, John. *Pictographs in the Interior of British Columbia*. Self-published, 1968.

Duff, Wilson. *The Indian History of British Columbia*. Victoria: BC Provincial Museum, 1964.

———. *The Upper Stalo Indians of the Fraser River of BC*. Victoria: BC Provincial Museum, 1952.

Eliade, Mircea. *Myth and Reality*. New York: Harper and Row, 1963.

Glavin, Terry. *Nemiah: The Unconquered Country*. Vancouver: New Star Books, 1992.

Harner, Michael. *The Way of the Shaman*. New York: Harper and Row, 1980.

Knight, Rolf. *Indians at Work*. Vancouver: New Star Books, 1978.

Laforet, Andrea, and Annie York. *Spuzzum*. Vancouver: UBC Press, 1998.

Neihardt, John G. *Black Elk Speaks*. Lincoln: University of Nebraska Press, 1961.

Teit, James. The Shuswap. Vol. 2 of *The Jesup North Pacific Expedition*, ed. Franz Boas. 1909. Reprint, New York: AMS Press, 1975.

———. "The Thompson Indians of British Columbia." Vol. 1 of *The Jesup North Pacific Expedition*, ed. Franz Boas. 1909. Reprint, New York: AMS Press, 1975.

York, Annie, Richard Daly and Chris Arnett. *They Write Their Dreams on the Rock Forever*. Vancouver: Talonbooks, 1996.

BRITISH COLUMBIA HISTORY

Harris, Bob. "The 1848 HBC Trail." In *The Best of BC's Hiking Trails*. Vancouver: Maclean Hunter, 1986.

Harris, Cole. *The Resettlement of British Columbia*. Vancouver: UBC Press, 1997.

Harris, R.C., and H. Hatfield. "Old Pack Trails in the Proposed Cascade Wilderness." Summerland, BC: Okanagan Similkameen Parks Society, 1980.

Hauka, Donald. *McGowan's War*. Vancouver: New Star Books, 2004.

Hill, Beth. *Sappers: The Royal Engineers in British Columbia*. Victoria: Horsdal and Schubart, 1987.

Koppel, Tom. *Kanaka*. North Vancouver: Whitecap Books, 1995.

Leslie, Susan, ed. *In the Western Mountains*. Sound Heritage vol. 8, no. 4. Victoria: Provincial Archives of British Columbia, 1980.

Lyons, C.P. *Milestones on the Mighty Fraser*. Rev. ed. 1956.

Reksten, Terry. *The Illustrated History of British Columbia*. Vancouver: Douglas and McIntyre, 2001.

Rothenburger, Mel. *The Wild McLeans*. Victoria: Orca Books, 1993.

CARIBOO–CHILCOTIN HISTORY

Alsager, Judy. *Gang Ranch: The Real Story*. Surrey, BC: Hancock House, 1994.

BC Parks: Cariboo District. *Churn Creek Protected Area Management Plan*. Victoria: Ministry of Environment, Lands and Parks, BC Parks Division, 2000.

Birchwater, Sage. *Chiwid*. Transmontanus 2. Vancouver: New Star Books, 1995.

Brown, James N.J. *Prospector's Trail: Poems*. Self-published, 1941.

Choate, Chilco. *Unfriendly Neighbours*. Prince George, BC: Caitlin Press, 1993.

Kind, Chris. *The Mighty Gang Ranch*. Self-published, 2003.

Knezevich, Fred. "Empire Valley Ranch." 1996. Photocopy of a paper in the Williams Lake Museum.

Koster, Henry. "A Brief History of the Empire Valley Ranch." n.d. Photocopy of a paper in the Clinton Museum, courtesy of Mike Brundage.

Logan, Don. *Dog Creek*. Self-published, 2007.

———. *Pioneer Pictures of Big Bar Mountain*. Self-published, 2005.

Marriott, Harry. *Cariboo Cowboy*. Victoria: Gray's Publishing Ltd, 1966.

Orchard, Imbert, interviewer. *Tales of the Ranches*. People in Landscape Series. Victoria: Provincial Archives of British Columbia, 1964.

Patenaude, Branwen. *Trails to Gold*. Victoria: Horsdal and Schubart, 1995.

Place, Hilary. *Dog Creek*. Surrey, BC: Heritage House, 1999.

St. Pierre, Paul. *Chilcotin Holiday*. Vancouver: Douglas and McIntyre, 1984.

Stangoe, Irene. *Cariboo-Chilcotin: Pioneer People and Places*. Surrey, BC: Heritage House, 1994.

Tatla Lake School Heritage Project. *Hoofprints in History*, vols. 1–8. Tatla Lake, BC: Tatla Lake Elementary Junior Secondary School, 1986–2000.

Ware, Reuben. *Five Issues Five Battlegrounds*. Chilliwack, BC: Coqualeetza Education Training Centre, 1983.

Witte Sisters [Veera Bonner]. *Chilcotin: Preserving Pioneer Memories*. Surrey, BC: Heritage House, 1995.

NATURAL HISTORY

Arno, Stephen F., and Ramona P. Hammerly. *Timberline*. Seattle: The Mountaineers, 1984.

Birchwater, Sage, and Ronald Cahoose. *'Ulkatchot'en: The People of Ulkatcho*. Anahim Lake, BC: The Ulkatcho Cultural Curriculum Committee.

Bridgeman, J.M. *Here in Hope: A Natural History*. Lantzville, BC: Oolichan Books, 2002.

Hebda, Richard, Nancy J. Turner, Sage Birchwater, Michèle Kay and the Ulkatcho Elders. *Ulkatcho Food and Medicine Plants*. Anahim Lake, BC: Ulkatcho Publishing, 1996.

Herrero, Stephen. *Bear Attacks: Their Causes and Avoidance*. Piscataway, NJ: Winchester Press, 1985.

Lyons, C.P., and W. Merilees. *Trees, Shrubs and Flowers to Know in British Columbia and Washington*. Vancouver: Lone Pine, 1995.

Parish, R., R. Coupe and D. Lloyd. *Plants of Southern Interior, British Columbia*. Vancouver: Lone Pine, 1996.

Russell, Andy. *Grizzly Country*. New York: Knopf, 1967.

Shelton, James Gary. *Bear Attacks*. Hagensborg, BC: Pogany Productions, 1998.

Shepard, Paul, and Barry Sanders. *The Sacred Paw*. New York: Viking, 1985.

Turner, Nancy J. *Food Plants of British Columbia Indians*. Part 2, *Interior People*. Victoria: BC Provincial Museum, 1978.

Turner, Nancy J., Laurence C. Thompson, Terry Thompson and Annie Z. York. *Thompson Ethnobotany*. Victoria: Royal British Columbia Museum, 1990.

Wooding, Frederick H. *Wild Mammals of Canada*. Toronto: McGraw-Hill Ryerson, 1982.

GENERAL LITERATURE

De Menil, A. and William Reid. *Out of the Silence*. New York: Harper and Row (published for Amon Carter Museum, Forth Worth), 1971.

LaChapelle, Dolores. *Earth Wisdom*. Silverton, CO: Finn Hill Arts, 1978.

Saner, Reg. *Reaching Keet Seel*. Salt Lake City: University of Utah Press, 1998.

Silko, Leslie Marmon. *Ceremony*. New York: New American Library, 1977.

Snyder, Gary. *Turtle Island*. New York: New Directions Books, 1974.

————. *The Practice of the Wild*. San Francisco: North Point Press, 1990.

Tobias, Michael, and Harold Drasdo. *The Mountain Spirit*. Woodstock, NY: Overlook Press, 1979.

van Der Post, Laurens. *The Heart of the Hunter*. London: Penguin Books, 1961.

OTHER RESOURCES

Kaiser, Henry, and David Lindley. *The Sweet Sunny North* [audiotape]. 1994.

Martel, Jan-Marie. *Bowl of Bone: Tales of the Syuwe.* [video] 1993.

Matthiessen, Peter. *Lost Man's River* [video]. 1990.

Copyright John Schreiber 2008

All rights reserved. No part of this work may be reproduced or used in any form or by any means — graphic, electronic, or mechanical — without the prior written permission of the publisher. Any request for photocopying or other reprographic copying must be sent in writing to Access Copyright.

NEW STAR BOOKS LTD.
107 - 3477 Commercial Street, Vancouver, BC V5N 4E8   CANADA
1574 Gulf Road, No. 1517, Point Roberts, WA 98281   USA
www.NewStarBooks.com   *info@NewStarBooks.com*

TRANSMONTANUS is edited by Terry Glavin. Editorial correspondence should be sent to 3813 Hobbs Street, Victoria, BC V8P 5C8 *terry.glavin@gmail.com*

Edited by Audrey McClellan
Cover by Mutasis.com
Cover photo by Marne St. Claire
Maps by Eric Leinberger
Typesetting by New Star Books
Printed & bound in Canada by Friesens
First printing, June 2008
Printed on 100% post-consumer recycled paper

The publisher acknowledges the financial support of the Government of Canada through the Canada Council and the Department of Canadian Heritage Book Publishing Industry Development Program, and of the Province of British Columbia through the British Columbia Arts Council and the Book Publishing Tax Credit.

LIBRARY AND ARCHIVES CANADA CATALOGUING IN PUBLICATION

Schreiber, John, 1941–
    Stranger Wycott's place / John Schreiber.

ISBN 978-1-55420-037-5
    1. Schreiber, John, 1941– — Travel — British Columbia   Chilcotin Forest Region. 2. Chilcotin Forest Region (BC) — Description and travel. 3. Chilcotin Forest Region (BC) — History.
I. Title.

FC3845.C445S34 2008   917.11'75   C2008-900072-2